THE DREAM OF
**ETERNAL LIFE**

# THE DREAM OF
# ETERNAL LIFE

BIOMEDICINE, AGING, AND IMMORTALITY

**Mark Benecke**

*Translated by Rachel Rubenstein*

COLUMBIA UNIVERSITY PRESS · NEW YORK

Columbia University Press

Publishers Since 1893

New York    Chichester, West Sussex

*Der Traum vom ewigen Leben: Die Biomedizin entschlüsselt das Geheimnis des Alterns*

Copyright © 1998 Rowohlt (Reinbek) and © 2002 Reclam (Leipzig)

Copyright © 2002 Columbia University Press

Library of Congress Cataloging-in-Publication Data

Benecke, Mark.

[Traum vom ewigen Leben. English]

The dream of eternal life : biomedicine, aging, and immortality / Mark Benecke ;

translated by Rachel Rubenstein.

p.    cm.

Includes bibliographical references and index.

ISBN 0-231-11672-1 (cloth : alk. paper)

1. Longevity.    2. Death.    3. Life.    4. Immortality.    5. Life sciences—Philosophy.    I. Title.

QP85 .B4413 2002

612.6'8—dc21                                                                                  2001047366

Columbia University Press books are printed on permanent and durable acid-free paper.

Printed in the United States of America

c 10 9 8 7 6 5 4 3 2 1

# CONTENTS

The laboratory where I research the genetic makeup of living human beings is separated only by a few doors from the chilly storage room where the Cologne University Institute for Forensic Medicine keeps its stash of cadavers. I was inspired to write this book by the daily existence, side by side, of different laboratories, one devoted to the blueprints for life and the other to the examination of tangible death. In my early years as a biology student, I wondered why nearly all living things die, despite periods of early development and the frequent lengthy existences of many life forms. Wouldn't immortality be a practical and worthwhile goal of evolution?

To search for an answer, I investigated strange reports of frozen heads, stored for regeneration—held in the vaults of private firms—with the same interest and intensity that I undertook straightforward experiments intended to isolate genes for aging. I met geneticists whose work entails the alteration of genetic codes and physicians who research ancient pieces of genetic sequences. I worked with developmental biologists who observe the origin and divisions of each individual cell of an animal through a microscope, and I found calculations that were meant ostensibly to explain the length of a person's life span. Among dusty mountains of books, I discovered works by the purported inventor of biorhythms and the teachings of Goethe's personal physician.

Common among these varied sources was the ancient, persistent human desire for eternal life. As long as humans have existed, we have used all trenchant and technical means to sustain and prolong life. Only through advances in biomedical research in the last few decades have we come closer to that desire.

Philip Gordon of Yale University seems to be sure of it: "In approximately the year 2050 we will have sufficient knowledge to be immortal." And in light of the international Human Genome Project (HUGO) (as well as Craig Venter's Celera Technologies), started between 1989 and 1991, the goal of which is the comprehensive mapping and sequencing of human genetic makeup (the human genome), it seems that the realization of this dream may be possible even before then. Some scientists have beat the drum for this inexorable march. The American molecular biologist and Nobel laureate James Watson compared the goals of HUGO with the attempts by the United States to be the first to land a man on the moon: "We used to believe that our fate was in the stars. Now we know it is in our genes."

The expectations generated by the decoding of the human genome are tremendous. Scientists have already localized the trigger genes for more than one thousand hereditary illnesses and have thereby found possible keys for their cures.

How are we to assess all this? The mundane manner in which modern biologists go about describing their work is astonishing, considering the fundamental existential questions Why do we live? Why do we die?

Up until this point, these questions have been mostly the domain of religion, philosophy, and literature. What can natural scientists contribute?

If we think of all the biological knowledge we have accumulated until now, we can observe the following basic principle: life as a whole is a stunning coincidence, a restless set of opportunities that constantly evolves in order to sustain itself—and will one day end where it began—in the icy cold grip of death. Life is its own self-purpose, as well as its own most extraordinary miracle. Its agenda is called adaptation, expansion, and multiplication. This book peeks into life's deck of cards to see how it answers questions of life and death. Unlike religion or philosophy, however, the answers will be presented in a scientific fashion—with the most simple and straightforward explanations.

*Cologne and New York*
*January 1998*

For the U.S. edition, the text was updated and revised. Molecular biology made the expected leap forward, and the reader might therefore enjoy our broad view on the topic of aging even more.

*Berlin and Chengdu*
*September 2001*

## ACKNOWLEDGMENTS

Maria Koettnitz, my German editor, fanned the small inspirational spark that grew into this book. Sebastian Vogel and Klaus Fehling provided both their creative and insightful assistance and referred me to critical source materials. Numerous antiquarian booksellers, among them Klaus Willbrand and Siegfried Unverzagt, sold me precious volumes from their collections that were otherwise impossible to find.

Many of my research colleagues, among them Mark van Thillo, Walter Doerfler, Maria Leptin, Klaus Rajewsky, Einhard Schierenberg, Svante Pääbo, and Michael Staak took the time in spite of their tremendously busy schedules to discuss and clarify many of the book's details. Olga Abendroth, Nancy Dise, Andreas Hantschk, Michael Hutter, John Marsden, Ganna Meleshko, Dorothy Munro, Mark Nelson, David Pescod, Wolfram Scheible, Matthias Schürfeld, Miguel Rodriquez, Marylin Ward, Andreas Wöhrmann, and Leonid Zhurnya each helped in their own way to shape this book.

Under the supervision of Holly Hodder, former publisher for the sciences at Columbia University Press, Rachel Rubenstein took the time and care to render the first and second drafts of the English translation. Linda Hotchkiss Mehta provided a thoughtful and meticulous technical review of the translation.

I thank them all for their excellent collaboration. Should any mistakes remain however, I shoulder full responsibility.

Let me state that this is a popular science book, and I am well aware of the legions of researchers working worldwide in the fields of biology, with many of whose works I take liberties in the discussions contained here. For the sake of clarity and space, I cannot mention every scientist and researcher, but I am indebted to them all.

THE DREAM OF
**ETERNAL LIFE**

## CHAPTER 1

## WHY DEATH IS PART OF LIFE

---

*A chain is no stronger than its weakest link, and life is after all a chain.*

WILLIAM JAMES

### LIGHTNING AND THUNDER

A muggy evening in the summer of 1952. In a University of Chicago laboratory, a charge of electricity, flashing like a bolt of lightning, charges through a flask filled with boiling water and a noxious mixture of methane, hydrogen, and ammonia gas. By creating these conditions, the young researcher Stanley Miller approximated the way in which his supervisor, Harold Urey, chemist and Nobel laureate, imagined the environment to have looked shortly before life appeared on earth. Miller almost created life in his witches' brew. What actually occurred was that a few chemical building blocks, amino acids, bonded together. Amino acids are the components of all proteins, the most essential substance of life. He was close.

In the beginning, there was mostly just reddish tarlike goo in Miller's flask. But, after a few changes in the setup of the experiment, the results grew far more interesting, so much so that they nearly confirmed a remark made in jest by the German biologist Ernst Haeckel at the beginning of the twentieth century: "Condense away—soon enough, something will start crawling."

Through their experiments, Miller and colleagues could retrace only the first step in the creation of life; further progress was simply impossible. What the U.S. scientist Harold Klein said still applies today: "Even the simplest bacterium is so vastly complicated from the perspective of a chemist, the way it came to be is simply unimaginable."

On the other hand, bacteria were certainly not the first lifelike creatures that developed on earth. This makes it somewhat easier to reflect upon the beginnings of life. Recent experiments by J. P. Ferris's research team at Rensselaer Polytechnic Institute in Troy, N.Y. demonstrated that the basic components of proteins (the amino acids) and the basic components of nucleic acid (nucleotides) can combine to form biological chains—under special conditions. These molecular chains appear in all living creatures and are thought to be the point of departure for life's development.

Miller's experiment demonstrated that life probably started a seemingly infinite number of years ago in the blackest of stormy nights.[1] With extraordinary effort, a colorful bouquet of life forms has blossomed since: buttercups, fir trees, crabs, eagles, poodles—and human beings. They all share one thing in common: a more or less impressive performance on the stages of life—and death.

## SMART CELLS

The more highly developed plants and animals are made up of an incredible number of biological components, commonly called cells. Some of these cells are just barely visible to the naked eye, but most are far too small. Cells can be so tiny that more than one million fit into a single drop of blood.

The figure of one million can be difficult to imagine. Try this thought experiment: an open hand can hold approximately two thousand grains of rice; both hands can hold about four thousand grains. Thus, 250 people's hands would be needed to hold million grains of rice: this is how many cells are contained in one drop of blood.

Not millions, but billions of cells work together systematically to make a complex organism like a human being develop. Regardless of how far the cells of a single body are spread, all their activities intrinsically coordinate with each other through an exact system that regulates what cells develop into which parts of the body. If that were not the case, each individual body part would perform its separate and distinct function randomly. A nerve cell, for example, could prompt the left eyelid to flutter continually. Or a small piece of skin, on the chin perhaps, might sweat, despite exposure to bitter cold. This is not the way in which a normal human body develops. An "agreement" that keeps the body functioning smoothly exists among all the body's cells. This profoundly complex system is the result of intricate chemical and electrical message systems.

There are about two hundred different types of cells in the human body, each with its own specialized task. Same-type cells have the same characteristic features: nerve cells are mostly elongated, while sweat gland cells are cup shaped. And each cell type's task corresponds to its shape: nerve cells conduct electric signals, sweat gland cells produce sweat.

Each cell contains "work instructions." These instructions, encoded in the cells, precisely dictate the cell's every activity. For example, nerve cells follow innate instructions to produce electric signals that control specific processes in the body, such as the blinking of eyelashes. If the nerve signal reaches its goal—the eyelid, in this case—the muscle cells contract according to their prescribed task and the eye blinks. The muscle cells cannot do otherwise. The precise function of a cell makes its action predetermined.

Do cells function like unflinching automatons that fulfill only a single prescribed task? Do they always carry out an encoded internal plan? Yes, but that's just the beginning. Almost all cells also contain a dormant treasure that ordinarily they do not use. This treasure consists of information for the entire living organism, its genome.

Each cell possesses not only the code for its own specialized task but also the set of instructions for every other cell contained within the living organism. Amazingly, the cell does not need most of the genomic information it carries. But regardless of where a cell is located in the body, its store of knowledge is equally as large as that of every other cell in the body. A human sweat gland cell carries the instructions for the production of veins, bones, and brain matter, although the sweat gland cell never produces these things. If it were merely a matter of indeterminate function, a nerve cell would produce fat just as well as the corresponding cells in the buttocks. It does not, however, and there are two reasons for this.

First of all, nothing tells a specialized cell to carry out more than its one function. There is no reason for it. Would a nerve cell in the brain take over the task of a fat-producing cell in the buttocks? Never. Second, specialized cells are positioned throughout the body such that any other function would be impossible. A cell walled into bone cannot release tears, and a fat cell cannot participate in thought. To do this, the cell would have to travel to the brain, elongate itself, and create connections to nerves: this is mostly impossible (research on the most basic kinds of cells, called stem cells, indicates that some kinds of stem cells can change into different kinds of cells when injected into a living organism). Why, however, don't cells discard their superfluous information blueprints?

Although the information seems superfluous, it could be tremendously valuable. It is an evolutionary remnant from ancient times. When life started about 4 billion years ago, life forms consisted of one single cell. Each single-celled organism needed comprehensive work instructions because it had to do everything itself. Although these organisms could find, gather, and use nourishment, they could not see or chew.

The fact that each cell carries all blueprints can still prove useful to humans from time to time. Several kinds of specialized cells read not only one particular work instruction—in accordance with certain environmental conditions—but also read "superfluous" instructions. When a wound heals and new tissue (usually skin and muscle) grows, such specialized cells go to work. They usually swim around in the blood and wait for a bodily accident or wound to happen. As soon as emergency nerve signals fire, they transport themselves to the exact destination via the bloodstream. Once there, they behave like an emergency doctor who pulls out the exact medical instrument from a first aid kit. Blood clotting agents and complex proteins that promote new skin and muscle growth in addition to pain relieving chemicals are but a few of the tools available to the "first aid" cells, called thrombocytes, leukocytes, and fibroblasts.


## THE GENETIC CODE OF LIFE

What does the cell's information, its blueprint, look like anyway? What is it made of? All information is found in long molecular strands contained in each cell. The instructions for aging and death are located somewhere along these strands. If one wants to be immortal, one has to alter a component on that information strand. To do that, one would have to know precisely how it is constructed and exactly where the individual instructions are located.

Sometimes the information strand is also referred to as one's genetic makeup or genome: the blueprint for an entire living thing. Physical characteristics, like the color of the eyes or the shape of the nose, can be transferred or passed down on the information strand from parents to children. The chemical name for this strand, which is really an acid, is deoxyribonucleic acid, or, DNA. Its chemical name sounds chemically exotic and magical, perhaps itself a remnant of some ancient culture. In translation, it means, "acidic sugar from the nucleus of a cell that has no oxygen (anoxic)."

DNA is an acid, similar to the kind in lemons or in vinegar, but its composition is far more complex. In lemon juice, the components of lemon acid swim in water. If one evaporates the water from the juice by heating it, for example, the components combine to form a lovely crystal. You could taste the sourness of those crystals, and they may glisten beautifully in the sun, but their structure is very simple, like basic geometry. No information can be stored or encoded in them because they consist solely of the same, identical crystal; the only variation is the size of the crystal. It's a bit like writing. One can place as many of the same letter in a row as one wants—a sentence will never come out of it.

The cell's DNA strands comprise four different kinds of acids. Information is encoded and stored there exactly the same way we create letter combinations in writing. The four acid building blocks of this information strand are arranged in a continually varying sequence: in groups of three, creating "words." A cellular translation apparatus reads each sequence and translates each one into specific cell components. There are sixty-four combinations, or words, some of which, however, carry the same meaning. All blueprints of the cells are encoded in this way, using this "language."

Each of the four chemical components, or "letters," is abbreviated like so: A for adenine, C for cytosine, G for guanine, and T for thymine. Assume that a tiny piece of the information strand consists of the order CCCGTTAAG. Put in highly oversimplified terms, the cell's reading apparatus recognizes that CCC = fat, GTT = on, AAG = foot, which prescribes, "Put fat on the foot."

Of course, the translated sequences of three do not actually form words as they appear in this book. Instead, cells use molecules as the basis of communication and their self-structuring. There is no essential difference, however. Fat really could be created, but several molecular detours would be necessary, and the cell would need a chain of at least ten thousand "letters" (approximately twenty pages of this book) to get that done. Most instruction sets are even longer.

Today, most biology students conduct simple experiments that translate instructions in DNA. First take a cell, then burst it, remove the DNA strand by pouring alcohol over it, and place it in a plastic test tube with saline solution. The contents of the test tube is then placed in a DNA sequencer, an apparatus about half the size of a washing machine. The sequencer recognizes the "letters" A, C, G, and T, and prints them out in the order they appear in the DNA. Place the test tube in the machine in the evening, and the next morning a page will be printed containing a vast number of Cs, Gs, As, and Ts, all arranged in their groups of three. Now, the sequence code can be read. Usually, however, this job is left to a computer program, which can translate the code more precisely and certainly more quickly than a person.

Although the DNA code of the cell is impressive, the idea behind it is even more so. Codes exist by virtue of the fact that not everyone understands them. The Gaelic word *líomóid* means "lemon." In Ireland practically anyone would know this; but since the Gaelic linguistic code is different from other languages, most non-Gaelic speakers would not recognize it. Decoding DNA is not impossible, however. But it does require precise work.

---

### How DNA is Extracted from Bananas

Somewhere in our genetic makeup lie the instructions for aging and death. If one were able to cut out the undesired instructions for aging from the corresponding strand of DNA, one might become immortal. To achieve that, however, one would first have to be able to find the right genetic strand. I have mapped out and successfully tried a simple method for extracting DNA. Here's how. One needs

- one-fourth of a ripe banana
- two and a half tablespoons of salt

(continued)

- a sharpened pencil
- a tablespoon of high-grade detergent (but not bleach).

Mash the quarter-banana with a fork. Spoon it into a glass, fill with tap water and add a tablespoon of high-grade detergent as well as two and a half heaping tablespoons of salt. Mix. The mixture can also be boiled for one minute—sometimes that helps speed the process. Using the pencil (with the tip touching the bottom of the glass), stir the mixture slowly clockwise.

After a short period of time, a small but easily detectable amount of a gelatinous substance will collect around the tip of the pencil. Pull the pencil tip along the inside of the glass and pick off the mass. Adding a good shot of ethanol to the mix just before will help separate the DNA from the proteins. The DNA becomes a more clearly visible to transparent substance and is now easier to remove from the solution.

You can fish around in the banana-salt cocktail repeatedly. Again and again, several thousand DNA-information strands together with as many proteins will wrap themselves around the tip of the pencil. If we were able to find and remove the death code on this DNA and reinsert the rest of the information strand back into a cell, an immortal banana tree might result.

(This experiment is as safe as the DNA itself. If you hadn't cooked and fished out the DNA, you would have eaten it with the rest of the banana.)

Where on the DNA thread can we find the information for aging? If I were to extract the DNA from a single cell in order to look for its aging instructions, I would be confronted with three problems: my fingers are too thick, my arms are too short, and my eyes are too weak.

The strand is so thin that it would be impossible to hold with my fingers. I wouldn't be able to see it for the same reason. Even the strongest magnifying glass is still too weak for this purpose. My arms are too short because the strand of a single cell is over six feet long—I wouldn't be able to pull it out in one piece. I have yet another, fourth physical shortcoming: my brain is too feeble. It is impossible to imagine that an acid-strand some six feet long is contained in a single cell, so small that it is invisible to the naked eye.

A similar but more serious hurdle, however, is that the sought-after information on the DNA strand is incredibly tiny. Stretch six feet of sewing thread across a table and imagine that this is DNA. Poke at it at some random point with a sewing needle. If the thread were DNA, you might have poked at the point of up to five blueprints for various cell components. These "blueprints" are also known as genes. A gene often comprises only one instruction, though sometimes an organism requires several genes for a single end product, say a beard hair or an eyelash.

Genes that belong together or resemble one another don't necessarily line up alongside each other. In this respect, they are similar to an encyclopedia of several volumes. You can place the volumes throughout your home as you wish. As long as you remember the volumes' various locations, you can still use the encyclopedia as a whole and look up any key words and cross-references you want. Similarly, each cell "knows" where all the related information it needs can be found on the DNA, and it can gather this information together.

Back to the DNA "thread" on the table. With the tip of a needle you have poked about five genes. (You were lucky, since more than 90 percent of DNA consists of what some biologists call "junk DNA," large sequences of DNA that do not seem to have any discernible function. Up until now these structures remain a mystery, but they will be deciphered in the years to come.) Imagine that each gene has a number: 1 to 5. Each number, each gene, is the blueprint for an ingredient in the anatomical recipe. All five ingredients together result in a useful whole, possibly the recipe for parts of a light-sensitive cell in the eye or for a tiny hollow sphere in a cell that transports molecules.

Each DNA strand contains several thousand various prescriptions. Imagine how often you could poke at that thread without touching the same prescription more than once. And, if you poke around enough, you might eventually hit the prescription for a mechanism that would allow "aging beyond the twenty-fifth year of life." But how can we find such a prescription in a real strand of DNA?

Using powerful microscopes, it is possible to see individual, fragile DNA strands. But the specific DNA sequences remain invisible. They are encased inside the strand of DNA, as if covered by a shell. And only the DNA shell can be discerned. What now?

A geneticist's adage goes: you can best find a gene by examining a living cell in which the gene you are looking for no longer functions. This sounds paradoxical, but it works: a cell that is different from all others is often easily recognizable. In order to find that elusive, ageless gene, you must wait to find a cell that does not appear to age.

Such cells do exist. These cells are immortal. But their appearance seldom occasions joy. To the contrary: immortal cells are usually cancer cells.

## IMMORTAL CELLS: HARBINGERS OF DEATH

In a family with a genetically inherited condition of four fingers on one hand, the DNA is altered at a single place in comparison with the DNA of other people. This is precisely the same with a nonaging cell: something has changed in its DNA. Such a change within the DNA-information strand is known as a mutation. Every blueprint of every cell can mutate. The DNA prescription for five fingers can change, just as the prescription for aging or anything else can.

A DNA change or mutation is neither good nor bad in and of itself. The genetically predetermined "loss" of a finger, for example, was the prerequisite for the development of hooves in horses, cows, giraffes, and camels.

There is little doubt that, if you are a horse, hooves are more useful (than fingers) for walking. The loss of aging, however, is advantageous for only a single cell. That cell multiplies itself rapidly and is actually similar in function to the primordial unicellular animals that were earth's earliest life forms. A cancer tumor is an accumulation of such immortal cells, almost identical to each other. Through the continual, uncontrolled division of one single mutated cell, the body finally loses its equilibrium. The cancerous growth might close veins, hinder the functioning of organs, and, in some cases, disfigure its victim.

There are many ways a normal cell can turn into a cancer cell. And there are many kinds of cancer. Because of the varied nature of tumors in both origin and appearance, there will never be one single treatment for all forms of cancer. There is also no preventive medication for it.

Cancer research has afforded major contributions, however: it has promoted extraordinary knowledge of the structure and processes in cells. Many biomedical advances would not have been possible without the contributions from cancer research. Since we now better understand cells, we can treat many other diseases that do not actually fall under the rubric of cancer. In addition, we have learned much about cells that may not have obvious practical application. This is the nature of basic scientific research: experiments intended to serve basic knowledge-seeking purposes often help other scientific areas. The solution to one long and intensively studied problem often comes from research focused on some other problem or investigation.

## THE PREORDAINED LIFE SPAN

One must know something about aging in order to understand death. It was once believed that cells die simply because they accumulate too much waste material after long periods of time. A cell's waste material is accumulated from breathing, digestion, and movement. These processes require energy, and energy produces waste. In the same way that spent nuclear rods are the waste product of nuclear power stations, so are particles of waste to cells. This waste material is encased within the cell in the same way used fuel rods are retained in nuclear power plants. Cells cannot dispose of noxious waste into an organism's body. For this reason, the cell most often retains the waste. It seems that, in this manner, the cell manages to poison itself slowly and, eventually, it dies. Philippus Theophrastus Paracelsus was supposed to have espoused this idea as early as the sixteenth century. Through the end of the nineteenth century it was nurtured by cell biologist Elias Metschnikow. Yet many biologists remained skeptical about the intended self-poisoning of cells. Another idea concerning cell death emerged at the end of the nineteenth century, by renowned biologists Charles Darwin and August Weismann, who assumed that a cell "wears out" like a machine. This comparison fails, however, when applied to a complete biological system. For example, muscle tissue atrophies when it is not used. When a bone is broken, the adjacent muscles are frequently immobilized and encased in a cast. This does not happen to machines.

A significantly more elegant explanation for cell death was discovered by scientists at the turn of the twentieth century, when an attempt was made to cultivate a blood cell in a glass petri dish. They placed a young, living cell in a drop of fluid, adjusted the temperature to a level conducive to growth, added nutrients and oxygen. The cell died. After that, the scientists improved the living conditions in the dish. They filled it with a gel consisting of boiled algae and placed the cell on top of it. Nothing happened. They added other nutrients on a random basis. They

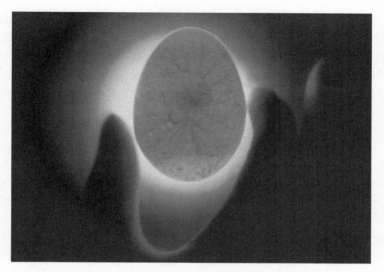

1 Hen egg in front of a light source. An animal's anatomy is planned down to the very last detail. Its blueprints are contained in the new creature's genes as well as in organic matter stored in the egg cell from the maternal organism. In this photo, we can see vessels of a developing hen embryo. (Photo: Bayer Inc.)

managed to get the cell to survive, but it did not develop further. Then, one researcher came up with the idea of adding blood serum to the nutrient culture. (Serum is that part of the blood that remains after blood cells and the fibrin clots are removed.) From then on, the cell thrived marvelously—it reproduced itself. A mouse cell kept in such a culture, for example, could divide itself up to twenty times. At first, the reason couldn't be explained. But it was noted that the cells divided more frequently with the addition of a 10 percent serum than with a 1 percent serum. The serum clearly contained the decisive secret of cell growth and cell division.

At the same time, in New York, surgeon, cell culture specialist and Nobel laureate Alexis Carrel and colleague Albert Ebeling, both of the Rockefeller Institute for Medical Research, found a way to keep connective tissue cells (culled from a chicken heart), called fibroblasts, alive in solution. (Chickens, particularly as unhatched embryos, are commonly used as research material, in part because chicken eggs can be easily incubated and studied.) Carrel was able to keep the fibroblast alive for a very long time. "On January 17, 1921," he reported to his colleague Raymond Pearl, "the connective tissue cells of the chicken heart will be nine years old." It took thirty-four years before Carrel's living fibroblasts were finally discarded (for no particular reason).

Carrel was initially concerned primarily with the storage rather than the cultivation of tissue. When he replaced a "finger-long piece" of a dog's stomach artery with a cat's vein of equal length, he first used tissue that had been frozen for twenty days. The operation succeeded. Carrel grew more daring. He knew of earlier work from 1894 in which transplanted skin grafts had been frozen for fifty days. Once the first organ transplants by researcher Carl Garrè were known, there was no stopping Carrel. Where he had succeeded with arteries, he now meant to succeed with significantly larger pieces of tissue. With another colleague, he finally reached a point where he successfully removed an animal's kidneys, ureter, and bladder and transplanted everything permanently into another animal. But he was still unsatisfied, even with this achievement. He convened with colleagues and improved the experimental apparatus established by American scientist R. G. Harrison, who in 1907 had sustained the growth of frog tissue in a nutrient solution. Harrison, at Yale University at the time, is considered to be the inventor of tissue cultivation, although Carrel's work is generally considered to be superior.

In 1913 Hermann Dekker described Carrel's cultivation procedure in the following manner:

> A small piece with a diameter of about one-tenth to one-half millimeter, let's say about the size of a pin-point, is removed from the tissue, is transferred to a glass slide and is covered with the prepared fresh plasma culture. This slide, culture side down, is cemented with paraffin to another glass slide, sandwiching the culture, and is then transferred to an incubator. The regrowth occurs. The entire procedure requires quick action; it would take only a few moments before the tissue could be contaminated by opportunistic germs. Since 1910, Carrel and Burrows have cultivated almost all adult tissues from dogs, cats, rats, rabbits, chickens—and human cancer cells.[2]

Within two years news about Carrel's experiments had spread to all major newspapers. In most cases, however, the experiments were misin-

terpreted or rendered falsely: pieces of tissue the size of pinpoints mor-
phed into complete arms and legs supposedly swimming in culture. No
one at the time recognized the true value of tissue cultivation, which,
one day, would be routine in cell research. "Scientists are such strange
fellows," Hermann Dekker wrote, "at first they don't care that their research has practical application. It suffices if in the solitude of their laboratories, their heads grow hot and their hearts are warmed by the
tremendous joy of the quiet successes of their work."[3]

Yet, these experiments continued to arouse great enthusiasm among
the general public and scientific community. Antoni Nemilow, professor
of anatomy and cell research of domestic animals, described Carrel's
work in 1927 from what was then Leningrad:

> The attempt to grow individual body parts outside of the organism actu-
> ally already implies a victory over death, for he (Carrel) proved clearly
> that science is stronger than death. Even if the little piece of flesh which
> has been living and growing in Carrel's laboratory for 15 years is very
> small, it has been snatched away from the jaws of death, which until
> now had been unconquerable and all-powerful. This growing, living tis-
> sue unconditionally contradicts once and for all, superstition and all
> fairy tales of the higher force of death, and stands above human will and
> reason.[4]

Although Carrel's experiment dealt with individual cells, not with a
distinct piece of flesh, one can relate to Nemilow's enthusiasm. Many
scientists at the time seemed captivated by the beauty of tissue cultures
and with the fact that it was possible to cultivate cells in dishes at all.

Between 1940 and 1960, tissue cultivation grew from the tinkerings
of individual scientists to full-scale operations. The smallest processes in
cells that formed the basis of life were now subject to real investigation.
The next challenge would be the systematic decoding of the cells' DNA.
Many laboratories began to cultivate cell cultures as source material for
their own research.

In the early 1960s, Professor Leonard Hayflick observed that connective tissue cells divide approximately fifty times in petri dish culture. As
the cells approached their fiftieth division, they divided more and more
slowly, until they finally died. Initially, the cells grew at a consistent

speed. As soon as normal cells covered the surface of the dish in a single layer, they stopped growing, but only temporarily. Cancer cells, on the other hand, continued to grow. Cancer cells have the ability to divide infinitely and would be the focus of cancer research from that point on. "Immortal cells," Professor Hayflick predicted at the time, "have one or several abnormal characteristics." An important insight resulted from this, namely, that if all abnormal characteristics could be discovered, then the origin of cancer would be understood. Perhaps then, preventive measures could be developed to avert cancer once and for all.

## CELL MEMORY

Hayflick went after cancer immediately. But he started with normal, noncancerous cells. In order to use cultured cells from the same source as long as possible, he selected one strand (called W1–38), which he then divided into several smaller pieces, freezing these in liquid nitrogen at -340°F. Starting in 1962, he sent samples of the frozen cells to colleagues all over the world (it is common practice among scientists to exchange samples free of charge). By the 1970s, the samples were the most thoroughly researched living cells anywhere—and Hayflick noticed something incredible. In spite of the icy, frozen slumber, the cells seemed to remember how often they had divided before they were frozen. Once defrosted, they carried out exactly as many divisions as they would have under normal conditions. The cells had some kind of recording mechanism that remembered the number of prior cell divisions. Although Hayflick may have suspected the workings of such an internal mechanism, this discovery was still a surprise. He described it in 1990, "We have defrosted cells from 130 samples in the last 28 years. Their memory is just as good as it was in 1962."

Cell memory functions similarly throughout every cell in a living body. If cells are extracted from an older person and placed in culture, the number of remaining cell divisions would be less than if the cells were harvested from a younger person. Nevertheless, a person's cells completely replace themselves several times in the course of a lifetime. At the moment of its inception, some mechanism must "inform" every cell just how old the rest of the body is.[5] And indeed, this is so.

Each cell in the body stems from another very similar cell and is transformed into a living organism through the exact regulation of gene expression. It can be assumed that the cellular information of a living body is passed on to subsequent generations of cells. What happens more intrinsically is still unknown. While several cell counting mechanisms have been discovered, cell biologists still don't understand exactly how developing cells achieve the precision of information transmission from cell to cell, function to function. For example, which mechanism is responsible for keeping count of cell divisions?

The older a person is, the less frequently his or her cells will divide in culture. Does this rule apply for all cells? As long as the cells are from a healthy body, the answer is yes. What happens, however, with cells samples from people who age early? Two diseases characterized by severely premature aging are progeria and Werner's syndrome. Children with progeria have the appearance of seventy year olds at the age of nine, while in patients with Werner's syndrome, physical decay, including hardening of the arteries, brittle bones, and a tendency toward diabetes, sets in just before puberty. The cells in the tissue of a person with Werner's syndrome divide a total of twenty to forty times. Hayflick's colleague Samuel Goldstein pursued this issue and put it to the test. He discovered that the Werner cells divided only eighteen times before they died. This experiment showed that aging among children afflicted with Werner's progressed even more rapidly than had been suspected. More has yet to be learned about this terrible genetic disease, and perhaps one day it will be eradicated.

At the beginning it seemed inconceivable that a cell's development could be dependent on one incredibly tiny internal mechanism, something probably contained in a few drops of blood serum. It turned out that only a handful of blood serum components influenced whether or not cells divided. Surprisingly, it was discovered that these components also function like growth factors or growth hormones.

Growth factors are closely connected to cell division because they act as signals—giving the OK for the complex processes that are actually completed before the cell divides and replicates. If a growth factor is missing, then cell growth falters. In that case, it no longer matters if the cell is fed nutrients or receives signals from other cells. If the correct growth factor signal is not transmitted, no other cell component will kick it into division mode. One can effectively maintain cells in a "growth factor-free" culture for weeks, stalling cell division. Even though the cells have all the nutrients they need to replicate, they remain in stasis. Without the correct signaling mechanisms that trigger growth and division, they exist in a kind of suspended animation.

Growth factors are proteins: each of them affects only certain types of cells, and each type of cell requires a precisely coordinated combination of growth factors. The body produces special growth factors at certain times over its life span and in certain locations throughout the body. When this occurs, those cells that are receptive to certain kinds of growth factors multiply and mutate. How much of a growth factor is needed? Consider the "sugar bowl" example. Throw a lump of sugar into the ocean and, assuming that it dissolves equally among all the oceans in the world, you could then take a cup of water from each ocean and find two molecules from that sugar lump in each cup of water. That rate of dispersal is a good scale by which to imagine the amount of growth factor used by the body.[6]

In one experiment, researchers tried to "coax" a young nerve cell in a glass dish to move as if lured by bait. A tiny amount of nerve cell growth factor was dripped in close proximity to the nerve cell. After a while, the cell grew a limblike extension and stretched itself toward the growth factor. If the growth factor was dripped onto a different location on the glass dish, the nerve cell followed.

Growth factors not only promote the development of cells but influence cells' activity. Cells will seek out growth factor. In fact, cells wither away if they don't find growth factor.

---

### A Deadly "Survival Cocktail"

Every molecular biologist or physician today can produce growth factors. Using bacteria, scientists cultivate the necessary proteins in culture. When the bacteria multiply, so do growth factor proteins. From this bacterial mass, scientists then isolate the desired growth factor "elixir."

If this bacterial cocktail guaranteed eternal life—a drink of life-enhancing growth factors—everyone might want a sip. Similarly, if nerve growth factor provided a universal cure for Alzheimer's disease, it would be an astounding medical breakthrough, worth fortunes.                                              (continued)

Unfortunately, such a cocktail, "even if the growth factor worked," would be poisonous. The indiscriminate use of growth factor would force cells to survive when they should not. Nerve cells that ordinarily do not belong together would grow together. Sight nerves might stimulate that part of the brain where hearing occurs so that a person might hear in color instead of hearing sound.

A "survival cocktail" of growth factors would never work, and would be lethal. The addition of growth factor proteins would throw the delicate balance of life-sustaining interactions wildly off kilter.

## THE MULTIPLE LIVES OF WORMS

If each cell dies at a predetermined time, then the death of an entire body can be foretold. In humans, however, the death of individual cells is difficult to determine. Human cells are exposed to many external factors during a life span, and these can alter the genetically preprogrammed "death date." Stress and smoking, for example, lead to an early death, while a healthy lifestyle can prolong life.

The life spans of certain life forms, however, can be directly observed and predicted. Such organisms live in the light of the world only a few days. One of these creatures is the nematode, threadworm (also known as roundworm or pinworm) *Caenorhabditis elegans* or *C. elegans* (see figure 2), and its time of death can be predicted.

These little worms can be found throughout the world. One finds them in flower pots as well as in fields and on farms. They are literally everywhere where there is soil, but they are barely noticeable because they are a mere half millimeter long and are mostly transparent. Biologists have taken pains to cultivate these creatures and to determine how long they live. They live about an average of twenty-one days.

Around 1990, certain laboratory strains of *C. elegans* worms turned up that, while exactly the same in appearance as others, lived approximately a half-day longer. *C. elegans* researchers all over the world discovered other worms that died after 12.5 or 33.25 days. Scientists' research instincts were aroused, and they began looking for a cause for the altered life spans. Although various explanations for altered life spans have been proposed, the factors that allow for such variations are still under investi-

Benecke 1994

2 The threadworm C. elegans, one of the first organisms to have its entire DNA
decoded. (Photo: Mark Benecke)

gation. One thing is clear: perhaps the secret of an extended or shortened life lies in the genetic substance in the form of "death genes."

## SEVERAL DEATH GENES ARE ALREADY KNOWN

Geneticists were able to find several genes in *C. elegans* that deliberately
kill cells within the worms' own bodies. Two particularly famous suicide
genes were named ced-3 and ced-4 and are found in every cell of the
threadworm.

As the young nematode grows, certain cells die off at foreseeable
times. This is not especially dramatic—the same process occurs when
human fingers form. Without certain programmed cell death, we would
have webbed hands and feet.

If both suicide genes are knocked out—which is possible with today's
technology—then all cells survive for a longer time, including those that
would ordinarily die during the normal developmental process. Particularly noteworthy are the so-called survival genes, such as gene p53. They

prevent programmed cell death without eliminating the death genes. U.S. developmental geneticists are already discussing the development of medications based on "survival genes." *C. elegans* researchers are very close to realizing this vision. Already in the early 1990s, Thomas Johnson of the University of Colorado succeeded in knocking out one death gene of the threadworm, thus doubling its life span.

A gene was recently discovered in another of biologists' preferred experimental organisms, the fruit fly (*Drosophila melanogaster*), and named "reaper" (after the grim reaper), since it introduces controlled cell death as soon as the specific number of death signals is amassed within a cell. If the reaper gene is eliminated, damaged or sick cells can continue living. This is far from optimal, since damaged cells must be eliminated absolutely or cancerous growths can occur.

Better candidates for the prevention of aging could be other death genes, as with the DNA segment coding for apolipoprotein E (apo E). Thomas Perl's research team at Harvard University discovered in the mid-1990s that very elderly people who naturally lacked the apo gene E4 lived very long lives. Of course, one can't be sure that there is a necessary link between a lack of apo E4 and old age. However, there is a close correlation between chronological onset of aging, which strictly speaking begins after the age of twenty-five, and our internal genetic programming.

## CELL SUICIDE

The hands on the clock of life consist of DNA, measuring out the exact age of the individual cells.

Yet many different aging processes are involved. One of these is known as "capping." Each end of the DNA molecule of individual chromosomes contains a "cap" of repetitions of the base sequence TTAGGG, which protects the DNA from "corrosive" substances. The cap actually resembles a knight's helmet that fends off the blows of a sword. Each time a cell divides and multiplies, the cap is temporarily removed and is quickly replaced upon completion of the cell division. In aging cells, the protective cap sometimes appears to be less durable with each successive division (it gets shorter). In very old cells, the cap even seems rather brittle. The cell may intentionally wear it down over time. Ultimately,

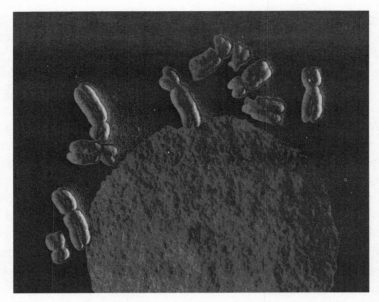

3 Human chromosomes enlarged three thousand times. They consist of a thin spiral-shaped acid strand (DNA). Individual pieces of the strand are made up of genes containing the complete blueprints for the body and its functions.

DNA-killing substances are able to deliver their deadly blows to the unprotected genetic substance (figure 3).

This idea also explains why old cells do not breed easily in culture. Beyond a precisely measured number of divisions, cells cease multiplying—the protective cap of the DNA had been destroyed. At this point, the addition of oxygen and growth factors would be useless.

But why do cells condemn themselves to death? There are three reasons. During development, growth of certain organs or limbs requires the "suicide" of other parts of the proto-organs or proto-limbs. This is similar to the process by which a sculptor forms a figure from a block of stone. Only by chiseling away shards of stone will the figure of the desired model emerge. It's quite the same with a young human embryo that still possesses web tissue. As the fingers and the toes grow, the areas of skin between them die away. This is genetically predetermined, and it makes good evolutionary sense. Once, our earth's earliest life forms needed webbed hands and feet, but over millions of years of evolution all that remains is a remnant that can be seen in developing embryos.

The second reason for intentional cell death lies in the fact that cells in high demand must be replaced regularly. This is the case, for example, with skin cells that are constantly exposed to the wear and tear of our environment. But even cells that comprise the human liver, calls that detoxify the blood, wear down. Sooner or later, they are replaced by new liver cells. All the cells in the body that are exhausted and cease to function are eliminated and replaced by new, fresh cells. In this way, nearly all of the body is renewed several times during the course of a life.

This general repair process, however, occurs at varied paces in various organs. Red blood cells are particularly short-lived: 2.4 million of them die each second in our bodies and are immediately replaced. Liver and bone cells, on the other hand, remain functional for several years.

### The Life Span of Some Cell Types

| Type of cell | Average life span |
| --- | --- |
| red blood cells | 120.0 days |
| lung cells | 81.1 days |
| ear skin | 34.5 days |
| lip skin | 14.7 days |
| stomach cells | 1.8 days |

During the course of one year, we replace:

228 small intestinal walls
193 stomach exits
25 skin coverings of the lips
18 livers
8 tracheas
6 bladders

After about seven years, we are new people in the truest sense of the word; only the nerves and muscles do not, to a great extent, regenerate. That we don't look as tender as newborns after seven years has to do with the fact that, first of all, not all cells are replaced simultaneously and, second, that the processes of aging and dying overlap with the regeneration processes. The constant, ongoing processes of aging and dying lead to the overall reduction of the rate of renewal and the materials needed for that renewal.

If the latter didn't exist, the human body could regenerate itself for a very long time—even eternally. Why then, must the arduously rebuilt organism die after seventy, eighty, ninety, one hundred years, or more?

The answer to this question serves as the third reason for the preprogrammed "voluntary" death found in the genetic substance of living things.

Even if it does not seem to make sense that we humans must die for the good of our species, this necessary component of survival—aging and death—is predetermined in our DNA. The reason, according to modern science, concerns a principle of evolution that transcends the individual: that of species adaptation to environmental changes. Individual immortals could not achieve this kind of adaptation, since they continually renew themselves from their own exact same genetic material. When the environment changes, only those descendants that fit better into the new environment than their ancestors—because of the inheritance of small coincidental mutations in their genetic makeup—will survive. Since no one can predict the changes that will occur—heat, cold, wind, toxins, ultraviolet radiation—descendants that inherit random adaptations are often the most successful. Parents pass on to make room for their children in the most literal sense, but their genes survive in some form throughout future generations.

This driving force of survival of the species—the creation and nurturing of slightly adapted descendants—far transcends the private interests of any one individual. For this reason, all multicellular life forms in the world today carry an overriding program that allows them to die and, before that, to produce children who though similar to them, are not exact replicas. This program is so strongly anchored in the genetic makeup that, generally speaking, it cannot be reversed.

The power of this fundamental force can be inferred from the phenomenon that many people who cannot have children nonetheless feel an urgent desire to do so. It is the "voice" of our genetic makeup that creates this drive, the so-called ticking of the biological clock. Yet there is never a rule without exceptions. In rare cases, some life forms achieve a kind of individual immortality.

## IMMORTAL CREATURES

The further biologists and geologists look into the past, the more amazed they are at what they find. At one time, the world was populated with strange looking animals and plants. In primeval oceans, for example, there was a kind of crab with terrifying horns that were as large as its entire body. Its back was covered with a few jagged edges and its gills jutted out from the sides of its body. Oversized, bent antennae probed its inhospitable environment. This long-extinct crustacean would not fit into a current ecological system. All the same, it lived 65 million years ago, which is recent, in geological terms. When these strange ocean creatures emerged, the dinosaurs had just become extinct.

An even more ancient creature that was neither plant nor animal lived in what is now Australia, China, and England. It measured over a yard in diameter and was almost flat: ten times as thin as a hand laid flatly on a table. The surface of this creature was lightly furrowed and crossing the midsection was a dividing line. That was it. On its smooth surface neither head nor antennae, eyes, mouth, nor tail could be detected. Perhaps this creature moved itself forward similarly to the way an earthworm does today—but as an enormous flat, wrinkled disk that pulled itself together and then extended itself. Perhaps, like the earthworm, muscle waves pulsed over its body. These wormy disks crawled around on the ocean floor more than 600 million years ago, well before there were dinosaurs or horned crabs.

Horned, or marrella, crabs and worm disks seem bizarre both in their appearance and way of life. But let's take a look even further back in history. In primeval times there were creatures that possessed a trait that was far more strange: they were practically immortal.

The eternal life of these animals contains a paradox, however. On the one hand, "eternal animals" do not die—unless a predator snatches them. On the other hand, however, such creatures are not one of a kind. Descendants of such immortal animals are direct copies of a shared mother. The mother may die, but the copied descendants are exact replicas of their parent: from the shape of the body to their reaction to a certain stimulant, a tremor or a flash of light. They are perfect mimetic copies of their predecessor, entirely indistinguishable from it. They are clones.

Survival artists like these still exist today. We encounter them daily without knowing it. One such example can be found in the drop of water from a pond, even without the help of a magnifying glass: the hydra. This multiarmed creature survives any injury by regenerating its destroyed or damaged body part.

The name of the animal comes from ancient Greek mythology. According to the Greek legend, the Hydra is a monster that grows several new heads for each head that is severed. This thousands-of-years-old legend contains the rudimentary observation that is plausible today. Even if these stories are hyperbolic, their descriptions still serve as a good basis for the naming of the actual animal. Reality often resembles legend: the hydra need not worry about a severed head—it simply grows another one. A hydra could even be cut into a random number of pieces without perishing. Out of each and every piece, a new hydra would emerge, or the pieces would rejoin to form a whole new hydra.

In stagnant ponds, these animals thrive unchecked. They reproduce by initially forming small buds on their stems. The buds develop into new hydras, detach themselves from the parent, and attach themselves firmly somewhere else. The parent itself does not survive over the ages as one individual. It survives in the form of multiple individuals in many locations.

The spherical colonies of the volvox genus could also be considered another type of immortal animal. Like the hydra, these creatures live in ponds and puddles. Since their bodies contain chlorophyll, they are able to nourish themselves with energy from the sun. Sometimes these hollow creatures multiply en masse and give the water they inhabit a green hue. In 1981, Jeffrey Pommerville and Gary Kochert, two botanists at the University of Georgia, discovered that each individual spherical creature dies a natural death after about four to seven days. Before that, they deposit "daughter spheres" inside their own hollow interiors. The "mother sphere" expires and releases her descendants, which then grow and form daughter spheres within themselves. The process can be continued interminably, unless some external catastrophe annihilates it. It seems, however, that even the volvox has an internal clock that runs out after several generations. The only salient difference between the immortality of the hydra and the volvox is that the hydra produces her daughters externally and the volvox creates them internally. Before the volvox, all living creatures would divide themselves into two new animals when they reproduced. No trace of the parent remained. The genus volvox was the first form of life by which the mother continued to live for a while, dying sometime later. Parents who disappear when dividing "into" their descendants, on the other hand, leave behind neither a dead shell nor a corpse.

Division into eternal life—with or without a corpse—produces clones, or precise replicas of the parent organism. If scientists or biological companies are someday able to produce identical people, they would repeat in principle what earth's earliest, simple organisms did, the process that nature tested out in the earliest hours of life. The difference is that nature adapted this reproductive process a very long time ago.

## ETERNAL LIFE HAS A DISTINCT DRAWBACK

During the course of the earth's millennia, many organisms were lost to evolutionary transience. And while many other "immortal creatures" still exist today—mostly bacteria—the more complex and advanced creatures are, the less they are prone to immortality. Why? If this puzzle can be solved, can eternal life be reinstated? To answer this, researchers took those immortal creatures from their "fountains of youth" and placed them under a magnifying glass.

Since the magnifying glass has become powerful enough—in the form of modern, high-powered microscopes—researchers have been able to observe "highly bizarre creatures, entirely invisible to the naked eye, whose survival is effected by miracle upon miracle. No severe coldness, no scorching heat can kill them, and a mere drop of water can bring them back to life after long periods of dessication." This description comes from the eighteenth century. What was observed were wheel-shaped organisms, called rotifers. Scientists were truly amazed to find that these tiny organisms could resurrect themselves, as it were, from dust. A pioneer in microscopy, Anton van Leeuwenhoek, dissolved a bit of dirt from his gutter in water and immediately observed what appeared to be the spontaneous generation of extremely tiny life forms.

For well over one hundred years, no one could explain where these creatures came from. Many researchers repeated the experiment with

4 Ostensibly the oldest tree in the world, a Huon pine stands on top of Mount Reid in Tasmania. Scientists estimate the enormous tree, or tree system, to be at least 10,500 years of age—perhaps even 30,000 to 40,000. (Photo: dpa)

dust that had been kept bone dry in hermetically sealed containers; they all got the same results. Life apparently "originated" in the dead matter almost immediately after it was watered.

These days schoolchildren take pleasure in dousing dried grasses with water and observing resurrected life forms after a few days. The emergence of these creatures doesn't astonish anyone anymore. Microscopic organisms remain in a dried state, wrapped in suspended animation, like "sleeping beauties" awaiting a drop of water to bring them back to life. Once this happens, hardly a moment passes before they begin eating and multiplying. This occurs in a rush, since a new dry period could happen again at any moment.

Today we know that single-celled animals don't simply appear out of water or thin air but rather crawl out of their sleeping capsules. In all probability, they are direct descendants of similar organisms that lived hundreds of millions of years ago. A parent cell grew at one point and divided into two. Both daughters nourished themselves, grew larger and likewise divided. From the four cells grew eight, and so on. Then a rest period may have occurred, forced by a drought or other catastrophe that killed almost all descendants of the parent cell. Some cells survived, however, and multiplied all the more quickly, once conditions were more conducive.

---

### Eternal Slumber

Single-celled animals are not the only organisms that can live in a state of deep sleep, or hibernation. Plants can too. The difference is that plants encapsulate their seeds only, while other animals' bodies fall completely into an extended state of dormant suspension—mostly when essential life nutrients grow scarce or disappear. The longest life span of plant seeds is no less amazing than the survival abilities of hibernating animals. The seeds of some plants can live longer than some human beings—as long as eighty years. Corn, onions, celery, and tobacco produce seeds that can live as long as half the common human life expectancy. The best known life spans of seeds that blossom after a dormant period are as follows:

| | |
|---|---|
| field clover | 100 years |
| potato | 200 years |
| lotus flowers | 250 years |
| buttercups | 600 years |
| field spark | 1,700 years |

Four-hundred-year-old cherry trees, 900-year-old beech trees, 1,900-year-old lime trees, and 4,600-year-old bristlecone pines make up woods and gardens worldwide. The oak, a symbol of gnarled perseverance, has been known to survive up to 1,300 years. These ripe ages are, of course, the exceptions. A "normal" oak tree would not grow much older than 200 to 300 years, a beech tree not much older than 140. Thus, plants often live much longer than their dormant seeds. Unlike seeds in slumber, which produce nothing, a living plant can produce several more germ cells.

---

Unicellular reproduction by way of self-division happens incredibly quickly. If a parent cell is healthy, it can produce more descendants than the entire human population of the earth. This is a condition of eternal life. Although the parent cell no longer exists as such, parts of it live on in all its children. It's exactly the same with the volvox and the hydra. Why, then, don't most of today's life forms reproduce in this quick and practical way?

Speed is not the only measure of success in reproduction. Depending on the life form and the living conditions, descendants will thrive with varying degrees of success. The parent cannot know how, necessarily, a future environment will look. Creatures that produce copies of themselves live on eternally in their identical descendants, but their children may not be able to thrive in a changed environment. Even a small change in outside temperature can force the metabolism of these animals to spiral into total chaos. (On the other hand, because there are so many of them, some offspring often survive many smaller changes in the environment. A high reproduction rate can ensure survival of a species.)

Many people are concerned about the extinction of animals as a result of changes in the environment. The death of species, however, is nothing out of the ordinary for biologists and earth scientists. Since there has been life on earth, many species have disappeared irrevocably, while other new species have emerged. Out of all species that ever existed—a total estimated between 5 and 50 billion—merely one out of every thousand original species exists today. However, the rate of extinction has increased enormously in recent years. Species that are dying out today cannot coexist within the environmental conditions created by humans.

Even before humans, there were time periods of tremendous environmental pressure, the result of which was the death of approximately 65 percent of all earth's species. These catastrophic periods are called mass extinctions.

One of these terrible catastrophes occurred some 560 million years ago, just prior to a period when life bloomed and radiated, in the Cambrian era. By some unknown cause, nearly all life on earth was wiped out and an entire world of organisms was destroyed in the boundary period between the Precambrian and the Cambrian ages. These early life forms, presumably neither fungus, nor plant, nor animal, are known to us only through Precambrian fossils preserved as stone imprints. In the fossils of this age we see "sea feathers" or "disk-like worm creatures," but, in reality, we know nothing conclusive about their lives or ecosystems.

During the Cambrian age, the precursors of contemporary organisms radiated extremely rapidly; hence the phrase *Cambrian radiation*. Coupled with periods of rapid expansions of life, mass extinctions periodically swept over the earth. The two largest occurred about 250 million years ago at the boundary of the Permian and Triassic ages, and the other more famous mass extinction occurred about 65 million years ago, when all the dinosaurs vanished.

David Raup, professor of geophysics at the University of Chicago, suspects that meteorites, asteroids, or bolides (brilliant meteors, often exploding) were the cause of the catastrophes. According to Raup and colleagues, geological mass extinctions can be documented at intervals of about 26 million years. When Berkeley physicist Luis Alvarez and his son, Berkeley geologist Walter Alvarez, first published their theory of mass extinction from meteorites in 1980, no one believed them. "It was," Raup reports, "as though someone had claimed that the dinosaurs had been shot by little green men from a spaceship." But initial amazement and disbelief were followed by serious discussion and worldwide exploration. Evidence of a meteor impact was discovered at the Cretaceous Tertiary boundary, which marks the period of time 65 million years ago, in the layers of earth's rock record, in locations worldwide. Finally, the location of the meteor impact was discovered and validated. Chicxulub, an undersea crater greater than sixty miles wide, was located at the Yucatan peninsula. Still, paleontologists and geologists disagree that the meteor impact was the sole cause of that mass extinction. There is abundant evidence that environmental conditions toward the end of the age of dinosaurs were very harsh. Mass extinction is not necessarily a catastrophe, however—it is also of biological value. In his book *Extinction: Bad Genes or Bad Luck* Raup advocates the idea that the driving force of Darwinian evolution, the gradual adaptation of new species to the environment, was and is possible only by virtue of mass extinctions that regularly make room for new, more fit species. Without periodic mass extinctions, Raup is convinced, the enormous variety of species on earth would never have happened.

For some living creatures, extreme environmental changes amount to crises or catastrophes, while for others they do not. During an ice age, for example, many animal species without inherited protection from the extreme cold die out, while others have such protection.

Many animal species that are exact replicas from a kinder, former world were particularly affected by such catastrophic environmental changes. Still other life forms survived and continue to survive environmental changes because they have evolved to engage in a different activity, namely, sex.

## THE EMERGENCE OF SEX

Sexual reproduction is essential to species' survival, and its sensual aspects for some species certainly may be little more than an evolutionary "trick" so that individuals keep at it, whether out of pure instinct or pleasure.

Biological progress by sexual reproduction occurs not only when the descendants are not perfectly identical to their parents but also when they differ from one another. In humans, this is readily observable. Among the myriad other creatures that inhabit the planet, this is not always immediately apparent.

Let's observe an incipient ice age: if the water temperature of an ocean sinks to about 50° F, then most living fish will probably freeze to death. Some types of fish, however, survive the duration of the cold. They possess a certain characteristic invisible to the naked eye: the ability to produce protection from cold, an internal antifreeze of sorts. The antifreeze you put into an automobile's radiator is a chemical fluid, most of which is alcohol derived. Fish antifreeze protection is derived from a protein, a "frost protection protein." Nature has actually produced fish that are entirely protected from freezing. They live in the iciest oceans of the world and produce descendants that differ from generation to generation. However, the favorable freeze-protecting trait is always inherited by their surviving children.

Sometimes, a few progeny of the "cold-resistant parent fish" possess either none of the "antifreeze" or they possess it in a different form. Wouldn't it be practical if all descendants were exactly the same, just as the descendants of the hydra or volvox? Then the inherited advantageous protection would not be lost.

But that would not be practical. The cold-resistant fish could indeed reproduce, by cloning themselves quickly for a period of time, since simple unicellular division is generally quicker than sexual reproduction. Yet eventually some other environmental change would be bound to happen, one for which the cloned animals had no protection, and such identical animals could then no longer make the necessary adaptations. They could all die in an instant. It is important that at least some descendants always be slightly different. Only in this manner can new traits and genes be tested—genes that, perhaps, will prove essential one day. This may also be the case for human beings.

Charles Darwin and Alfred Russell Wallace were the first to recognize these correlations and connections. On July 1, 1858, Darwin's theory of evolution was read before the Linnean Society of London. The theory rested on two findings: that species tend to produce slightly varying versions of themselves in their descendants and that the modified descendants develop further over the course of generations through the selection for environmentally advantageous traits. Each new life form of a species can add something new to the further development of that species. The further development of life, the adaptation to altered environmental conditions, and the colonization of new living spaces is made possible solely by the differences among individual members of a single species.

Sexual activity is not a prerequisite for this: mutations can also result from asexual cell division. The main advantage of sexuality lies in the fact that new combinations of various inherited traits can be combined and tested. Adaptation to the environment accelerates itself in this manner, especially among animals that have only a few descendants. And some animals make use of more than one method of sexual reproduction. For example, water fleas shift gears temporarily at times of adverse environmental conditions from self-reproduction and cloning to partnered sexual reproduction. Some single-celled bacteria mix their genetic makeup as a suitable means of safeguarding and preparing for the future. Their seemingly simple sexual activity consists of exchanging bits of DNA with similar bacterium over cell-to-cell bridges created on the spot. Later they multiply themselves, and their offspring express the changed genetic information. Once a new bacterium begins to form, however, its genetic makeup is immutable; no changes can be made. Their fundamental genetic traits are now unalterably established.

In order to equip species with sufficient genetic variety, there are multicellular organisms (threadworms and some snails) that become hermaphrodites and function as both male and female partners, producing egg cells as well as sperm. The male-female process is employed only in emergencies and never for routine self-reproduction. One exception does exist: when environmental conditions are clearly excellent, some species (hydras and volvox) produce many identical descendants very rapidly. However, as soon as environmental conditions grow noticeably worse, these tiny life forms return to more normal levels of sexual reproduction. The original meaning of sex is simple: survival of the species through preventive adaptation to a changing environment.

## CHAPTER 2

## NO ONE WANTS TO DIE

---

*We all labour against our own cure, for death is the cure of all diseases.*

*SIR THOMAS BROWNE*

### I SAW A LIGHT AT THE END OF THE TUNNEL

Many people firmly believe that life continues after death. There even seems to be proof of this. People who have narrowly escaped death tell stories of images that seem to show that death is merely the gateway to an unknown future. Usually they describe a "dark tunnel," at the end of which shines a light. The reports are amazingly similar: "I felt light and content. As I moved toward the alluring light, I felt timeless and weightless. I saw the blurred outline of a person. Just before reaching my destination, I heard my name. When I came to, I was lying in a sick bed and was looking at the faces of my loved ones. I could not and did not want to speak. I still carry that kernel of happiness and peace that I experienced in those minutes on my way to the hereafter. My life has changed. I am no longer afraid of dying."

Such accounts are not new. The Roman attorney Apuleius (born around A.D. 125) described a similar experience in his novel *The Golden Ass:* "I went to the point of death. At midnight I saw the sun in light streaks; I walked before all the gods, from face to face, and prayed to them in the closest possible proximity."

Another commonly described scenario is this one: "I felt myself separating from my body. Slowly, my soul ascended, and I could look down upon my own body. I saw everything as if through a veil; I could even recognize my loved ones who were standing beside me. I was not astonished or amazed, but alienated in a strange way. Everything was different, but somehow also the same."

Experiences of this kind do not seem commensurate with the worldview shaped by the natural sciences. In fact, scientists and physicians explain such experiences without recourse to faith in the supernatural. They have confidence in an interpretation of such events that seems reasonable, in physical, testable circumstances. (This, of course, suffices only until someone comes up with a more convincing thesis.)

One thing is clear from the outset: it is entirely impossible that separate individuals could have the same strange experience in different cultures and under completely different circumstances. Whether these people are religious, atheist, pompous, imaginative, shy, or none of the above, they invariably report on at least one of the two death motifs, either the tunnel or the out-of-the-body experience.

That all these stories have been invented is impossible. There must be some sort of process that is the same or at least similar among all people who have come close to dying. We know that the body creates biochemical substances, as they are needed, in every conceivable situation. When you touch your nose, the nerve endings release chemicals, especially hormones, that enable the brain to combine the messages "Nose touches finger" with "Finger touches nose." If you glimpse a sexually attractive person, your endocrine system releases hormones that cause your heart to beat more quickly and alter your thoughts. If you lose something valuable, the interplay of many other chemicals, like adrenaline, induce anxiety as you search for whatever was lost. The effect of hormones on human behavioral responses has been thoroughly researched. Some hormones are even used as medication. Some behave like "narcotics" similar to morphine. Therefore, it is entirely possible that the body releases hormones in our moments closest to death, and these hormones may play on our nervous systems so that they help to conjure up the images so often and repetitively described. These death motifs, one might imagine, are a kind of thought pattern already inscribed, chemically or hormonally, in the human body. They need only be retrieved, especially in extreme circumstances.

It is even feasible that such a hormonal, chemical mechanism is hereditary (and passed on from generation to generation). Two reasons suggest this explanation. First, we are already familiar with thought patterns related to our faculties of perception: we see a cat, and we say to ourselves that we see a cat, for example. The structure of the brain, as it develops from the earliest stages of life, constantly reinforces and retains certain images in the creation of its own database. For example, we still see the white page of a book as white even if we are wearing tinted sunglasses, because we "know" the nature of white pages from when we first encountered a white, printed page. This is because our brain has learned to cull accurate information from "false" data.[1] Similarly, our brain perceives the quickly sketched outline of a human body as such. The brain nerves essentially "add" missing pieces of the sketch to form a complete image. In other words, the brain informs us in an evaluative or even "partisan" way about the external world. There is much we don't see and much that gets added in.

In the 1960s, David Hubel and Torsten Wiesel, two Harvard neurobiologists, radicalized the accepted notions of brain development by showing that experience itself could alter the development of the brain—even after birth. During their twenty-year collaboration, they demonstrated this by focusing their work on fixed visual perceptions of monkeys and cats. They monitored the brains of the laboratory animals by placing electrodes onto their brains (through tiny holes drilled in the skulls) and determined that the animals elicited a regular response to different visual stimuli: straight lines, circles, moving objects. They found that seeing out of both eyes was necessary for the normal arrangement of so-called ocular dominance columns, and that these orderly columns of brain cells responded to visual activity from one eye or the other. They discovered that these ocular columns changed in response to external, visual cues—that the brain was "plastic" and capable of change in response to the environment.

Since the basic structure of cats' and monkeys' brains differs only slightly from the human brain, it can be concluded that human brains, like these animals' brains, function automatically, even unconsciously, as we take in the world around us.

It could be said that the Harvard research team solved, to a great degree, the mystery of the soul. When Hubel found out in 1981 that he and his colleague Wiesel were to receive the Nobel Prize for medicine, he said, "It was always assumed that a brain cannot understand itself, or rather, cannot understand the consciousness contained therein, for that would be like pulling oneself by one's own hair out of the water. We think this is nonsense. One can study a brain as well as one can a kidney."[2]

The amazement at the discovery of these scientists was tremendous and long lasting. The way had been paved for the notion that humans do not think entirely freely but can, at least in some situations, be steered by patterns of thought.

There is a second reason that speaks to this mechanism: there are actual chemical hormones, called endorphins, that are produced by the body when it is hurt, stressed, or nears death. The production of endorphins leads to a total relaxation of the muscles. This is why mice, once caught, hang limply in the jaws of their feline predators.

The pleasant floating feeling described by some who have narrowly escaped death stems from the same muscle-relaxing endorphins produced by the body. Even the feelings of happiness and contentment that have been felt by nearly all people who have "returned" from death could possibly be a result of the effects of such substances. We try as often as possible, whether consciously or not, to place ourselves in situations where our bodies will produce endorphins and other pleasure-producing hormones for pleasure, from moments of rest and sex to moments of stress. It isn't for naught that we refer to work-obsessed people as workaholics. It could be the endorphin "rush."

When the body nearing death releases such "drugs' into the bloodstream to suppress pain or to relax the musculature and the mind, the total cooperation of mind and body takes place for the last time. Such a death is peaceful and mild.

It is nonetheless entirely unclear why humans do this. Perhaps death is only one of several states of stress to which the body responds in a similar fashion.

## "WHAT KIND OF SLEEP IS IT THAT SEIZES YOU NOW?"

People of very early cultures did not understand the process of death as we do today. The only definition presented so far has been that of the natural sciences: the body and brain no longer function.

Presumably, though, our ancestors perceived the processes of death and dying in their own ways. These people had probably not developed a culture of self-consciousness that provided for consideration of their own deaths as we do. This has been inferred from a text that was discovered one hundred years ago from an archaeological site that was once known as the city of Nineveh, on the Tigris River. In the library of the Assyrian king Ashurbanipal, who lived around 650 B.C., were unearthed stone tablets that tell the epic of Gilgamesh, which includes the story of the great flood. The content of this story is astonishing and allows for far-reaching conclusions about the relationship between an ancient culture and the process of death.

Kurt Aram, who studied very old manuscripts at the beginning of the twentieth century, tells us the story of Gilgamesh, the king of Uruk:[3]

The myth is concerned primarily with the death of Gilgamesh, who, when confronted for the first time, with the death of his friend Enkidu, makes futile efforts to escape this same, incomprehensible fate. The experience marks Gilgamesh so thoroughly that it serves as the focal point of the myth and remains so throughout the entire story.

> Gilgamesh, stunned, stands before his dead friend:
> "What kind of sleep is it that seizes you now?
> You look dark and you don't hear me."
> —Yet he no longer opens his eyes.
> Gilgamesh feels his heart, but it beats no more.
> He then covers his friend like a bride.
> Like a lion he roars,
> Like a lioness who has been robbed of her cubs,
> He turns to the dead one,
> He tears at his hair.

Enkidu, Gilgamesh's dearest friend, was younger. When Enkidu lived, he had dreams and premonitions of his own death that Gilgamesh simply never understood. Once Enkidu died, Gilgamesh searched for his ancestor Utnapishtim, who was supposedly immortal. By finding him, Gilgamesh sought to avoid death.

> When Utnapishtim spies Gilgamesh across the sea, he says to himself,
> Why is there someone sailing in a ship that does not belong to me?
> He who comes is not a man,
> He does not have the right hand of a man?

Utnapishtim can offer no consolation to Gilgamesh, when he returns to his homeland, Uruk. Still unsatisfied, he searches for answers in the underworld. He asks Ea, the only benevolent god among the ancient Babylonian gods, to conjure up Enkidu's spirit from the underworld. The following dialogue develops between Gilgamesh and Enkidu's spirit:

> *"Pray tell, my friend, pray tell, my friend,*
> *What is the nature of the underworld? Pray tell."*
> *"I cannot tell you, my friend, I cannot tell you,*
> *If I wanted to tell you the nature of the underworld that I have seen,*
> *You would be compelled to sit the whole day weeping."*
> *"Then I will sit the whole day weeping."*
> *"See the body you once touched, the body which gladdened your heart,*
> *It is being eaten by worms like an old dress,*
> *My body, which you touched, which gladdened your heart,*
> *Has disappeared, has turned to dust.*
> *It has crouched down in dust,*
> *It has crouched down in dust.*
> *If we ask the meaning of this, there is no remedy for death."*

Kurt Aram, however, refused to be satisfied with this interpretation. The Gilgamesh epic is ancient, at least dating back as far as 2700 B.C. But what if the story is much older? One archaeologist, Edgar Dacqué, seems to think so. In any event, these early human communities interacted with their natural environment and interpreted natural causes far differently than we do. To us, they appear to lack logic and reason.

In fact, ancient people had a magical understanding of nature. On this subject, Kurt Aram writes,

Of course people died natural deaths then as they do today, and they comprehended it more than a simple honey bee would comprehend death. Ancient people grieved for their dead when they perceived the phenomenon of death in their communities, but they did not *reflect* on death. They did not reflect on the condition of death as though in a state of sleep without awakening, or on the fact that worms feed on corpses like they would feed on an old dress. For them, no one really died—they passed on to another place, an underworld, and the spirits of the dead could be conjured up. Gilgamesh never comprehends that his friend Enkidu will not return, that he is absolutely gone, that he is dead. Rather, he pleads with his immortal ancestor for a means to avoid death.

Otherwise he would not have undertaken the long, dangerous journey to reach Utnapishtim, at the outer limits of the world and over the waters of death, which continuously threaten precisely what he wants to save, namely his life.[4]

This ancient legend resembles a dream in many ways. Dreams arise, so it is said among artists today, "when the cerebrum is tired." For all our brains are so highly developed, there are parts of the brain yet, like our earliest ancestors, that slumber still. The cerebrum often suppresses messages sent from these deeper layers of the brain. This explains why we understand death medically today and have some level of control over it. At the same time, we are terrified of this mysterious phenomenon just as our ancient ancestors were, especially afraid of being eaten by worms (actually, maggots) "like an old dress." We still yearn to know what happens after we die.

## THE BODY PRESERVED FOR THE AFTERLIFE

Ancient Egypt's cult of the dead demonstrates just how old these fears are. The Egyptians believed life after death to be entirely self-evident. They mummified corpses to give the deceased a protective shell for their bodies. In the afterlife, they believed, people received a new body. At funerals, they sang "fortune has arrived." Yet survivors still mourned the loss of their loved ones.

From the 2050–1570 B.C. on, the model for mummifiers and participating priests was Osiris. Commanded by the sun god Re, Anubis, the god of the dead, descended from heaven and prepared Osiris's body for resurrection. The soul of the mummified man was then loosely attached to his body. Nachtmin, the protector and head of the granary, wanted to "let myself pass continually on the banks of my pond, let my soul flutter in the trees I planted, let myself cool under the sycamore trees" after his death.

The mummifying of the ancient Egyptians was carried out in the following way. The entrails of the deceased person were removed and were dried with salt and sodium bicarbonate—since salts dehydrate bodily tissues. The entrails were then stored in special jars. Resins, oils, and aromatic substances served to embalm the rest of the body. Finally, various

5 A mummy can from the German Pharmacy Museum in Heidelberg dated 1825. The incredibly good condition of Egyptian mummies led European doctors from the Middle Ages through the mid nineteenth century to sell *Mumia vera aegyptica,* i.e., mummy fragments, as medication. In some pharmacies, mummy powder was available as recently as the beginning of World War II. (© Benecke)

objects were placed in the body, for example small linen packets, wax figures, and scarabs—beetles used as talismans and symbols of resurrection. Wax amulets sealed the incisions in the body, and gold coating protected the lips, tongue, fingers, and toes. The corpse was then wrapped countless times in linen, sealed again with amulets. In order to make the mummies look as lifelike as possible, they were adorned with a mask and painted boards. Herodotus wrote that a complete embalming took an average of seventy days.

This labor-intensive undertaking is a very good way of retaining decisive DNA. DNA is not particularly sensitive to dehydration as long as other destructive elements do not pervade its environment.

The amazingly pristine condition of Egyptian mummies led European doctors from the middle ages to the nineteenth century to sell *Mumia vera aegyptia*, little pieces of mummy, as medication. In some pharmacies, mummy powder was available up until the beginning of World War II (figure 5).

6 A tree burial in New Guinea. In the Western world, people are buried in wood coffins—even when bodies are cremated, the coffin is burned along with it. Other cultures burn their dead on funeral pyres, and in New Guinea, corpses are buried in deleafed treetops. (Photo: Wulf Schiefenhövel)

Like the Egyptians, every culture has its traditions for handling the dead. In parts of New Guinea, the dead are even "buried" in the tops of trees that have been stripped of vegetation specifically for this purpose (figure 6). Wulf Schiefenhövel described such a "tree burial." This physician, anthropologist, and longtime assistant of the behavioral scientist Irenäus Eibl-Eibesfeld lived in the highlands of Irian Jaya in New Guinea for two years. There, in the village of Munggona, among approximately 180 people of the Eipo tribe, he observed the following.

Ebna, a twenty-two-year-old from the Eipo tribe, died in the afternoon of June 2, 1975. Early the next morning, about five hundred yards from the center of the village, "several men and boys" began to prepare a burial tree. First they built a kind of scaffolding and removed the branches with leaves. At the top of the tree, they placed a chairlike burial frame. "Kaberob, the leader particularly for the sacral traditions of village life, urged the villagers to the funeral," Schiefenhövel reported. "The deceased was pulled up in stages by the men who prepared the tree and placed in the burial frame so that he faced the mountainous region of Mangedelo—that is where, according to Eipo tradition, the spirits of the Mekdumanang clan dwell, to which Ebna belonged. The body was then bound in this position and is surrounded by a covering of ferns and other leaves (*Pandanus adinobotrys*) so that it was concealed entirely. Remaining branches of the burial tree are cut and broken off." Four days later, on June 7, members of the tribe removed the covering of leaves, removed the body from the frame and tried to raise it in a net to the top of the tree. This failed, so they laid it at the foot of the tree, wrapped it in the net, and let Ebna's brother carry the body into his garden. There they lifted the body up into a new tree for reburial. "All those," Schiefenhövel writes, "who came in contact with the body, which has already shown significant signs of decay, rubbed their hands and body parts after this reburial with the leaves of stinging nettles." About a year later, Ebna was buried yet again. Ebna's brother and several helpers brought the mummified corpse into a small house where it was placed in a wood coffin. There, the remains decomposed finally to a skeleton. Sometimes, after complete decomposition, the Eipo bring their dead to the mountain cliffs, where skulls and bones find their final resting place.

## THE UNDEAD

While the dead Eipo is provided finally with a resting place, many myths describe the plight of people who are unable to die—although they long to. One example of this is the story of Count Dracula. This legendary figure is certainly not exclusively based on the historical ruler Vlad, who viciously beat back the Ottoman armies in the fifteenth century. Moreover, even the subliminal sexual implications of the vampire stories of the Victorian era do not explain the original reason for the legend—though they certainly explain its popularity.

The notion of vampires who hunt their victims at night and return to their coffins by daybreak harks back to a significantly more realistic, if a bit less romantic, phenomenon. This phenomenon can be observed today by only a very few, since contemporary culture avoids nearly all contact with corpses. In earlier times, corpses were almost always laid out at room temperature so that family and friends could bid farewell to the deceased. Before the body was laid out, undertakers or relatives prepared the dead body: depending on its condition, the body was combed, touched up with cosmetics, and dressed. If the tendons had contracted and the limbs were bent, they were stretched out again, sometimes with tremendous force and weighed down with a stone. The mouth, often open, was closed and a book was placed under the chin. Before the final viewing, these props were removed. It was frequently observed that limbs contracted again immediately or that the mouth fell open again "like a vampire." The physician Christoph Hufeland was witness to these processes. More frequently than not, these phenomena led to a total re-examination of the corpse. This was done not out of superstition or fear of ghosts but to make sure that the body was, in fact, dead.

The fear of "apparent death" was particularly widespread in the nine-teenth century. Entire books were written on how to bring such victims back to life. The advice ranged from "blowing air into the mouth" to "opening a vein" to "wrapping the victim in mustard with a Spanish fly." During such attempts, the body had to be moved a bit, and, on occa-sion, bloody fluid flowed from the deceased person's mouth; even today this observation is made at some autopsies when corpses are positioned so that the doctor may examine for stab wounds in the back.[5]

David Dolphin, professor of chemistry at the University of British Columbia, offers another possible explanation for the invention of blood-thirsty vampires. According to Dolphin, there was a hereditary disease that appeared in the late Middle Ages in large numbers among the east-ern European nobility, who were the originators of the vampire myths. Specific altered genes were consistently grouped together through the widespread marriages among close relatives, thus amplifying the spread of the illness. This illness was known as porphyria, or *Porphyria erythro-poetica,* and causes the upper lip to recede and the skin to become se-verely chapped and bloody. These symptoms are greatly exacerbated when the sick person is exposed to sunlight. Physicians in the Middle Ages apparently advised the patient to stay in darkened rooms (in their castles) and to compensate for blood loss by drinking animal blood. Centuries later, victims of porphyria were often treated by bloodletting. To replace the lost fluids, these poor people were required to drink cow's blood like their fellow sufferers in the Middle Ages (figure 7).

7 *Les buveurs de sang.* Certain traits of the disease porphyria are similar to characteristics attributed to vampires. Through the nineteenth century, people with this illness, depicted here in a slaughterhouse, were prescribed cow's blood. (Drawing: Olga Abendroth, based on a painting by J. F. Gueldry from the nineteenth century)

A painting by J. F. Gueldry (figure 7) from the nineteenth century called *Les buveurs de sang* ("The Blood Drinkers") depicts the suffering of porphyria victims. It shows a butcher taking warm blood from the neck of a cow, tied down by a helper, then filling a cup with the blood and giving it to the first victim in a row of six or seven sick people. The expression of disgust and horror but also surrender to fate is utterly heartrending. The fact that porphyria victims are pale and still have red lips and teeth, though their teeth grow differently, makes it understandable that these people were taken once for vampires.

In 1997 the English researcher David Pescod-Taylor compiled a list in the periodical *Bizarre* of the "vampiric" traits that can appear in dead people. Pescod-Taylor worked under the assumption that a range of skin diseases and so-called delayed death appearances such as the rotting, detachment, and dehydration of skin, the loss of teeth, individually or together, led to such fallacies.

Even today, some people would take a lip-smacking corpse, posed in different positions in a coffin and with blood like liquids running from its mouth, to be a vampire. It is also understandable how wealthy, unapproachable nobility who drank blood with pursed lips might have caused people to be fearful and suspicious. The physician Christoph Hufeland was one of the first, around 1800, to dispel this myth.

The actual vampire myth became known to the German-speaking world when, in 1733, reports were spread from the village of Servia (on what was then the Turkish border) that the dead rose from their graves at night and sucked the blood from the living. Literally all the German regions of Europe were struck by panic and feverishly updated by the local newspapers of the time. Ultimately, the Prussian emperor had some corpses in question excavated from their graves. Many were impaled, beheaded, and—for good measure—burned. Michael Ranft, who was a witness to these events, was absolutely convinced that these ostensible vampires were actually *Scheintote* (people who had been buried alive— cases of people who had merely been taken for dead—*scheintot* in German literally means "seemingly dead")

In 1734 he described the fate of these people with great sympathy:

> And if the person who was taken for dead and then reawakens was a loving father, if he escapes his tomb using every last bit of strength and manages to stumble to his family on weak legs, if this helpless man then realizes that his sudden return from the graveyard causes nothing but sheer horror, death, and destruction among his loved ones—won't he collapse again, struck with the same horror, once again become death's prey out of pain and grief, helpless and abandoned?

The literary figure of Count Dracula originated in Bram Stoker's novel *Dracula*, written in 1897. Two books written on a rainy night in 1817 on Lake Geneva served as a model for Stoker's novel: Mary Shelley's *Frankenstein* (Shelley was twenty years old at the time) and Polidori's *The Vampyre*. Bram Stoker's version, however, is linked to the prince of Walachia, Prince Vlad the Fourth (Vlad the Impaler, Vlad Tepes, or Kaziklu Bey). Vlad lived in the fifteenth century and was famous for his successful and incredibly cruel battles against the Ottomans. After he had spent his youth as a slave in the Ottoman territory, he ruled from 1448 to 1476, during which were intermittent periods of exile. His knowledge of his enemies had apparently made him all the more successful in war.

Vlad's father (Vlad the Third) was a *dracul*, a knight of the order of the Dragon, an order of the Holy Roman Empire. In the vernacular, because of an error in translation, he was known as "the devil." If one accepts the family moniker, cruelty was obviously a family tradition.

Since the ending -*a* means "son of," Vlad the Fourth was named *Dracula,* son of Dracul. Where his remains might be buried is unknown. The stories of Dracula's actions persisted in Romania down to Bram Stoker's day. Inspired by a history book, the author used the more or less true historical background for his vampire story and gave the original father of all literary vampires a name: Dracula.

An article from the daily newspaper *Express* from the Rhineland is testimony to the fact that belief in Dracula is astonishingly widespread (or promoted, at least, by local tourist offices). On May 4, 1996, it ran the headline, "Dracula Is At It Again," and reported, "For some, they are simple drops of moisture on a stone, for others the face of Dracula grins from the castle wall. The site of the apparition is a castle in Sighisoara, Romania, where the Walachean Prince Vlad Tepes lived for a while."

Even in the German-speaking world there are reports of people who behave like vampires. Fritz Haarmann and Peter Kürten, two of the most notorious German serial killers, quenched their thirst for blood on the necks of their victims. Both committed their murders into the beginning of the twentieth century. Kürten was mesmerized by the "rush of blood" from his victims, whom he stabbed to death. In 1930 in Düsseldorf, he decapitated a swan and drank its blood with relish. Fritz Haarmann continuously maintained that he "bit his victims [boys] to death" in a fit of passion. These claims could neither be proven nor contradicted, since Haarmann cut up his victims and disposed of them.

Newspapers like the *Abendzeitung* of Munich use the catchword *Dracula* as well and reported, "A 75-year-old man in Berlin was taking a walk with his dachshund Jockel, when a 43-year-old man yelled, "I am Dracula!" and bit him in the neck. The obviously intoxicated attacker was apprehended. One hour after the dog owner was assaulted, he died—of a heart attack" (April 6, 1995).

In all literary versions of the Dracula story, the burden of eternal life and endless wandering is most prevalent; never-ending life is what makes the count such a sad figure. And Dracula's case is hardly unique. Most of the old stories of the undead are permeated by melancholy and suffering—the Flying Dutchman (in European legend a captain and his ghostly ship doomed to sail the seas forever) and Frankenstein's monster are two such examples. It is no coincidence that all creators of the un-dead depict them as eternally punished. For what else would eternal life be besides repetition, emptiness, fatigue, and boredom? Writers make a certainty out of what is biophysically very probable: if death does not curb life, life loses its value.

---

**Real Vampires**

There is one kind of vampire that exists in reality. They do not live in Eastern Europe, however, and they have the zoological nickname "little bats." Three of the vampire bat species indeed nourish themselves by sucking blood. At night, the four-inch common vampire, *Desmodus rotundus,* on all fours, creeps up on its victims (usually hoofed mammals) and cuts open their skins with razor-sharp teeth. It then sticks out its tongue as far as possible and rolls it down and to the side into the shape of a tube. Through this "tube," it spits saliva into the wound that contains a substance to prevent coagulation and then sucks out some blood. The common vampire bat only occasionally requires a blood meal of one hour. This does not cause the host animal much harm; the vampire however can distribute diseases such as rabies.

Other namesakes of the so-called true vampires are the bat species Cuban flower vampire and the Jamaica fruit vampire, but there are more. One thing they share in common: they possess some of the fastest metabolisms of any living mammals and they need to feed very frequently.

---

## WHEN DOES AGING BEGIN?

There is much truth in the idea that a person is as old as he or she feels. Mental health and agility are necessary to feel vital. The genetic, biological system by which the body ages runs its course more or less uninfluenced by the environment and by a person's emotional state.

The first phase of life, from birth until about the age of twenty-five, is the development phase of the organism. Afterward, the aging genes begin to hinder the production of new bodily substances. From that point on, the body maintains essential functions without producing much new bodily material.

For this reason, a grown person can survive for several weeks without nourishment. Extensive fasting in the early years can lead to permanent damage because constant nourishment is needed for the development of organs and other body parts. Later in life, the need for nourishment decreases, and, as long as liquids are consumed, basic bodily systems can be maintained for quite some time. It even seems that living creatures have the ability to nourish themselves less and less the older they become. Whether humans beings do this is another question. The initial fasting months of anorexia victims or the long fasting periods among certain religious groups proves that the necessity of nourishment for the maintenance of the most basic bodily functions is far lower than is normally believed.

Biologically speaking, from the time of puberty until the age of about fifty is the most critical period in a person's life, since it is in this phase that descendants can be produced.

Actual physical aging begins to become noticeable around the age of fifty. Women around this age most frequently enter menopause, after which childbearing is no longer an option—although modern medicine is tinkering with that option. Men, on the other hand, can produce sperm until they are quite old. Charlie Chaplin serves as proof: he fathered a child at the age of seventy.

Nonetheless, the body's metabolism changes significantly after the age of fifty. This is the cause of many of the widely known signs of aging.

As his or her short-term memory weakens, an aging person has a harder time learning new things. It takes longer and longer for nerve connections necessary for learning to be produced. Other body functions like the production of hair pigment stop somewhat later; the result is white, colorless hair. The reverse happens with age spots. In this case, more pigment is stored in the skin. Skin cells produced at an advanced age continually decrease in biological quality, and the natural biological guards that prevent the production of such cells in young people fall asleep—and wake up less and less often.

The body starts to shrink, and skin becomes more wrinkled. Most severely affected by shrinking are those areas of the body that were not renewed often during the course of life. Among these are the eye lenses as well as the vertebra discs of the spine. The fact that some elderly people are hunchbacked, however, is not necessarily a true sign of aging, but it is more prevalent among the aged. This condition arises from illnesses, most often osteoporosis, and is noticeable among its victims primarily as they age, since the disease itself develops slowly and is itself a part of the aging process.

---

### The Course of Life

People who research age read the developmental path that courses through each individual human life. There are characteristic traits and developmental focus points for each phase of life, as the following table sketches:[6]

| | |
|---|---|
| **babyhood**<br>until 18 months | movement; simple language;<br>total dependence on others |
| **early childhood**<br>until age 7 | well-developed language;<br>perception of gender difference; group<br>games; preparation for youth |
| **late childhood**<br>ages 6–13 | good reasoning ability (but still slower than adults);<br>team sports |
| **youth**<br>ages 13–20 | puberty; independence from parents;<br>sexual relationships |
| **young adult**<br>ages 20–44 | profession and family |
| **middle age**<br>ages 45–65 | professional goals reached; time of self-reflection<br>"empty-nest" and mid-life crisis;<br>retirement |
| **old age**<br>ages 65 until death | satisfaction of accomplished life goals;<br>dependence on others; loss of partner;<br>poor health |

This list would be no more valuable than a bad horoscope if it didn't provide specific predictions about the course of life. Such predictions do exist. The most well-known example is the series of phases that comprise the development of the ability to walk, which every mother and father observes in their children. Often, the exact month can be predicted for when a certain crawling or walking ability will be reached. In addition, the exact time a child learns to grasp an object can be predicted with relative success.

Countless cell deaths in the course of life precede the death of the entire organism. The death of cells, organs, and tissue is altogether different from the death of an entire being, as demonstrated in the example given in part 1 of dying cells in the development of our hands. In any event, an organism can survive the death or loss of individual body parts, such as the spleen, a leg, or even a quart of blood. This explains why we can now say that humans already begin aging at age twenty-five, even though the actual aging process begins much later: from twenty-five on, bodily functions are slowly lost.

Nonetheless, an individual organ, such as a donated kidney, can outlive the donor. Frozen blood can be transfused to another living body years after the death of the blood donor. Hence, it is not the concept of aging that is so difficult to describe, but the concept of death. The heart could still be beating in a person considered "clinically dead." Among other reasons, death has to be defined with great precision in order to make organ removal possible. When physicians and legal scholars refer to death, they mean the same thing meant instinctively by everyone today in the Western world: the irreversible damage of the entire brain.

I will come back to the issue of organ removal and the contemporary debate surrounding it. But first let's take a look at the practice of medicine with the living.

## THE HUMAN BODY'S FOUNTAIN OF YOUTH

By selectively manipulating a strand of DNA, researchers may one day find a way to stop the aging process. This possibility is reminiscent of the age-old dream of a fountain of youth from which one drinks to become young again.

Of course, it's questionable whether a fountain of youth would really be a good thing. But if we step back for a moment and take a look at life as a whole, we realize that the human species achieved immortality long ago. We must simply broaden our field of vision to regard the eternal youth of humanity rather than the eternal youth of individual human beings. The genetic information carried by each individual contains extremely old components. Some of the information is much older than the individual; some of it was even contained in the first creatures that ever lived on earth. Nearly all animals possess such DNA segments, even if small in number. These can be blueprints for important, fundamental organs, limbs, and more. But even entire patterns of behavior—for example, the escape response, as in "flight or fight"—based on the same DNA information, could be the same among flies, foxes, and humans. As long as there is life, this information will not disappear anytime soon, since it is so critical for species' survival. Consequently, we call it immortal.

But we weren't speaking about eternal DNA, but the entire life of an entire people. Here the individual pieces of DNA are hard at work, using the same immortality trick: a piece of the parents survives in every baby. This is seen in the facial characteristics of newborns that most often resemble their fathers and mothers. Character traits are also hereditary, particularly when parents either encourage or discourage certain talents or behaviors. This human form of immortality is really no different than the fountain of youth. Through the "real" fountain of youth, individuals continue to live as younger members of the entire family of humans. What lives on in real people is the appearance and the characteristics of their ancestors.

After the classical idea of the fountain of youth—life without aging—and the notion that many parental characteristics live on in their children, there is a third concept of immortality. The human body consists of two main parts. One is the part we refer to as the "body"—the head, chest, hands, and so on. The second part consists only of those areas of the body that are immediately linked to the production of germ cells (gametes), that is, the sperm cells produced in the male testicles and the egg cells produced in the female ovaries.

Some biologists, for example England's Richard Dawkins, consider adult bodies to be simple vessels carrying the germ cells along the genetic path from one generation to the next. A well-known example is the male praying mantis, which is sometimes eaten, head first, by his female partner after successful sexual reproduction. With the passing on of sperm, its life purpose is complete; from this moment on, its body is useful only as nourishment to the pregnant female.

However, American evolutionary biologist Steven Jay Gould asserts that the often-cited cannibalism of female animals is more the exception than the rule. Approximately fifteen years ago, Gould compiled all known original publications on the subject and discovered that female praying mantises have actually been observed eating their male partners only very few times. When Gould published his study, he inspired other researchers to observe the behavior of praying mantises. These experiments led to the conclusion that male praying mantises only seldom become the prey of the females.

There is relatively little information even about the most famous partner-eater, the black widow. And the following description of the love dance of tarantulas by Germany's well-known animal expert of the 1980s, Vitus Dröscher, was not proven until very recently: "The male jumps around his bride so wildly out of fear of being devoured. Usually, it manages to escape."

Other animals, such as the female desert scorpion, *Paruoctonum mesaenis,* indiscriminately attack any creature that approaches and is small enough, including the small males of its own species; the males are killed and eaten just as often as any other small animal.

In spite of the incomplete information about the reproductive behavior of animals, the example of the praying mantis and the other animals mentioned here is rather informative. It shows that a male creature loses its basic right to life after successful fertilization, unless the male participates in the raising of the offspring.

John Alcock, a zoologist from Arizona, described the role played by the reproductive cells and what function the rest of a living creature's body has. Some butterflies manage to escape attacks from bats through skillful flying maneuvers that manage to astound both biologists and flight experts. Alcock concluded his description of these maneuvers with the words "If the butterfly is successful, it effectively postpones its inevitable fate and thus gains additional time to reproduce." Seen in this light, even the most impressive aviation tactics of these butterflies are executed for their (not yet extant) descendants. Only one thing is important for natural selection in all these cases: the immortality of the gametes and their genes. Mortal individuals are little more than pawns within this system.


## GOETHE, HUFELAND, AND A MISUNDERSTANDING

Christoph Wilhelm Hufeland wrote one of the first books on the subject of nourishment and life span. *Macrobiotics; or, The Art of Prolonging Life* was completed in 1796 and is still widely known today. Hufeland lived from 1762 to 1836 and was an extraordinary physician who did not treat his patients on the basis of diagnostic assumptions and handed-down beliefs, but stood on the threshold of scientifically proven medicine. In Prussia, he completely reformed health services, pushed through the smallpox vaccination, and introduced the thermometer as a basic doctor's instrument. In addition, Hufeland was Goethe's primary doctor for some time—perhaps the following remark made by Goethe refers to him: "We live as long as God sees fit, but there is an enormous difference between living pitifully in old age like an old dog and living fully and in good health. A clever doctor has great influence in this respect."

Contrary to the misunderstanding that Hufeland's title allows for, his book is not a handbook for immortality. It is concerned more with the means of so-called macrobiotic dietetic methods for living as long a life as possible, within nature's limits. The basic principles of macrobiotics are simple and sound remarkably modern; Hufeland describes them like this:

Human life is, physically seen, a strange animal-chemical system, a phenomenon that causes the competition of unified forces of nature and ever-changing matter. This system must have its own operative laws, limits, and duration, insofar as it depends on the measure of the energy and matter available, its use of this energy and matter, as well as other external and internal conditions. But it can, as any other physical system, be enhanced or hindered, stalled or accelerated . . . Here, dietary and medical rules that extend the length of life can be formulated; indeed, a science can be made of this, known as Macrobiotics, or the art of long life. Macrobiotics must not be confused with medicine or medicinal dietetics—these have other means and other limits. The purpose of medicine is health, the purpose of macrobiotics is long life. . . . Thus, practical medicine is to be regarded as an instrument in relation to macrobiotics. Medicine teaches us to recognize, prevent, and cure diseases, which make up part of life's enemies, but it must be subordinated to the higher laws of macrobiotics.[7]

What were macrobiotics, in their clear distinction from medicine, really like in Hufeland's time? The stated rules were simple and would appear banal to people today: the rules of macrobiotics recommend a peaceful, balanced lifestyle with enough but not too much sleep, general hygiene and skin care, plenty of exercise outdoors, and psychological calm. Hufeland also considered a well-rounded diet important, not least of all as a way to maintain healthy teeth. While all this advice seems to stem from Hufeland's daily experiences, another recommendation was most certainly based on a moral code: Hufeland advised against sexual acts in youth and outside of marriage. He regarded marriage, however, as the locus of "domestic and public happiness." He counted masturbation, as well as "the living together of people in cities" (due to poor air quality) to have damaging, life-shortening effects. Boredom, an overabundance of activity, and overeating were also on his list of taboos. Hufeland was absolutely right in his recognition of how critical a well-balanced, low-stress lifestyle is. He also warned against unnecessary anxiety about death—a piece of advice that is perhaps even more appropriate today than it was in his time—if one considers that the discovery of penicillin and other antibiotics during World War II raised the average life expectancy in Western countries by ten years. Since then, devastating epidemics such as the plague and typhus, which had altered the course of history time and again, have been nearly eradicated.

Along with his astoundingly modern views, Hufeland also represented other views that have been forgotten entirely by contemporary civilization. It is worth taking a short look at them, since, first, they created a basis for Hufeland's work, and, second, they are another example of the kind of confusion that can be caused by nonscientific work.

Despite the advances in the natural sciences at the beginning of the nineteenth century, people believed—until the pioneering work done by Charles Darwin (*On the Origin of Species*, 1859) and by the physician Claude Bernard (*Introduction to the Study of Experimental Medicine*, 1865)—in a vague "life energy" (*vis vitalis*) that was found only in living creatures and made these creatures live. At the beginning of the twentieth century, this idea was taken up again by "neo-vitalists" and led to a bitter fight among scientists. The second-generation vitalists not only assumed the existence of an unknown life energy but also ascribed responsibility for the progressive development of life (as seen by humans) to this force. This blatantly contradicted the research results of Alfred Russel Wallace and Charles Darwin, who had shown that the development of life could be explained with the physical laws of nature and especially with observable biological data. In 1914 Ernst Haeckel, the first significant German biologist to be convinced by Darwin's theory of evolution, railed against the vitalists as he represented their theory:

> In very clear and disconcerting opposition to the mechanical progress of modern biology, a mystical direction known as "neovitalism," also known as "palavitalism," which tries to bring new validity to long-buried superstitions of old myths of supernatural life-energies, has made a name for itself in the course of the last two decades. Without revealing any new facts in support of this, confused neovitalists have attempted to legitimize the ostensible "autonomy of life" and of organic processes through a sophisticated dialectic. The fact that it has nonetheless achieved some level of esteem is explained by the regrettable increase in confusion that has occurred among many modern scientists as a result of short-sighted specialization on the one hand and an inability to judge general conditions philosophically on the other.[8]

Hufeland, who lived a century earlier, was a first-generation vitalist, just as most of his contemporaries were. "Without a doubt," he wrote, "the force of life belongs to the most general, incomprehensible, and powerful forces of nature. It fulfills, it moves everything, it is most probably the spring from whence all other energies of the physical or at least organic world come. It is that which brings forth, maintains, rejuvenates everything; it is responsible for the creation of every spring that blooms with the same splendor and freshness as the first one that came from the Creator's hand many thousands of years ago."

From what today is recognized clearly as a false conception of life energy, Hufeland nevertheless developed modern health recommendations that are still used in contemporary life. Through a healthy lifestyle, "life energy" and, by extension, the body is strengthened.

Hufeland's influence endured through the twentieth century, manifested in the health care system in East Germany. In the 1980s, East Berlin's Museum of Hygiene sponsored an exhibition in Erfurt with the title "Health Is Fun." To demonstrate the link between a healthy lifestyle and aging, five East German researchers developed a test by which the visitors of this exhibition could determine their "biological age." The test consisted of several questions about weight, gender, and body type; in addition, the researchers measured blood pressure, hearing ability, and

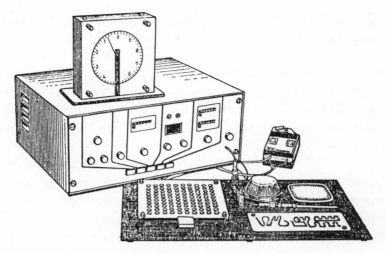

The "Geromat" (from the Greek *geron* for "old man"), manufactured in the 1970s at the university and hospital in the then Eastern German town of Halberstadt, takes both mental and physical characteristics into account when measuring biological age.

dental condition. The result: nearly everyone who took the test was biologically younger than their "real" age suggested. (The research team explains this, however, by noting that the people who visited the exhibit were particularly active both in mind and body.) A group of scientists from Leipzig had already developed a test apparatus in 1971 that could be used to determine biological age. The basis for the construction of this machine was research conducted on rats by Alfred Kments and Gerhard Hofeckers at the University of Vienna. It was called the Geromat (from the Greek work *geron*, for "old man") and was finally completed by the University of Halberstadt and Halberstadt Hospital.

The Geromat takes both mental and physical characteristics into account: blood pressure, dental condition, hand strength, and other features such as reflexes, memory, and coordination. Forty-seven individual points are tested, and the test takes one and a half hours. With such tests, the age of a body *at the time* can be rather precisely determined. The Geromat, however, does not necessarily predict future physical well-being.

With another, more simple, test, the prospective age an individual will reach in his or her lifetime can be calculated—assuming that external conditions do not change. This so-called age table was developed with the help of thousands of scientific observations. The underlying studies are concerned with those factors that alter life expectancy. As many life conditions as possible that measurably influence age are taken into account.

## THE LIFE TABLE

This table, which is actually a kind of survey, can calculate a person's age, if there is no accident to cause an untimely death.

Surveys of this kind are already used in the United States by life insurance companies to determine risk factors and, by extension, to determine premiums. The test is only accurate on average. As Marc McCutcheon describes in his book *The Compass in Your Nose and Other Astonishing Facts About Humans*, it could certainly be the case that "your life expectancy comes rather close to the calculated result [of the test]."9

Here is the test:

Begin with the number 74 and add or subtract as follows:

- if you are male: -2
- if you are female: +4
- if you live in an urban area with more than 2 million inhabitants: -2
- if you live in a rural region with fewer than 10,000 inhabitants: +2
- if all your grandparents lived beyond 85: +2
- if one of your parents died before the age of 50 of a stroke or heart attack: -4
- if a close relative has or had cancer or heart disease under the age of 50, or has had diabetes since childhood: -3
- if you earn more than $40,000 per year: -2
- if you have completed high school: +1
- if you achieved a higher educational degree: +2
- if you live with a partner: +5
- if not: -1 for every ten years over the age of 25 without a partner
- if you work at a desk: -3
- if your work is physically stressful: +2

- if you exercise 3 to 5 times per week for at least 30 minutes: +4
- if you exercise twice a week: +2
- if you sleep more than 10 hours per night: -4
- if you are often aggressive, tense, and angry: -3
- if you are generally relaxed: +3
- if you are usually in a good mood: +1
- if you are usually unhappy: -2
- if you smoke 20 cigarettes a day: -7
- if you smoke 40 cigarettes a day: -8
- if you smoke 10 cigarettes a day: -3
- if you consume more than 45 grams of pure alcohol a day: -1
- if you are 10–30 pounds overweight: -2
- if you are 30–50 pounds overweight: -8
- if you are checked annually for cancer: +2
- if you are 30–40 years of age: +2
- if you are 40–50 years of age: +3
- if you are 50–70 years of age: +4
- if you are over 70: +5

Add a bonus point to your result for each of the following that applies to you:

- blood pressure under 130/75
- cholesterol under 200
- resting pulse under 60 beats per minute
- no difficulty breathing or asthma
- no chronic illnesses
- if you live with a house pet
- if you are working beyond the age of 62
- if you are a light eater
- if you eat breakfast regularly
- if you have friends besides your spouse/partner

Subtract a point if any of the following apply to you:

- blood pressure over 140/90
- cholesterol over 200
- if it takes a long time for you to recover from physical exertion

- if you are anemic
- if you are sick more frequently than average for your age
- if you get out of breath easily
- if your resting pulse is over 60 beats per minute
- if you do not eat breakfast regularly
- if you have no friends besides your spouse/partner

(Anyone who wishes to read the amount of time one has in one's lifetime can do so, from data such as these, with the electronic life clock, certified as an invention in the United States since July 9, 1991, under the patent number 5031161 for David Kendrick. Instead of the hour of the day, this wristwatch shows one's remaining years, hours, minutes, and seconds.)

Of course, neither breakfast nor a pet nor frequent good moods have a direct, life-lengthening effect. Rather, they reflect generally positive environmental factors that often cannot be measured more precisely. The *exact* way a house pet improves an older, solitary person's quality of life is as inexplicable as the reason a child dies when she or he lacks human physical contact.


## A BRIGHT CANDLE BURNS MORE QUICKLY

Human beings are warm. They are soft. They move, and blood flows in their veins. A fingernail, once ripped to the cuticle, grows again, and two pints of blood are replaced completely within a few weeks. These processes consume energy that we get from food and drink. Small "motors" exist in all our bodily cells that convert nourishment into available, usable energy.

Fifty years ago, scientists were already trying to measure how much energy a person uses in his or her lifetime. Particularly during the first twenty years of life, these energy researchers determined, people require a tremendous amount of energy. This is to be expected, of course, since a growing body must build tissue.

During the course of the rest of one's life, the need for nourishment or energy decreases. This too is not surprising. Day by day, physical and mental functions such as memory or flexibility weaken with increasing age—slowly but surely.

The most crucial measurement for those interested in eternal life is the "internal energy candle." From birth to death, men consume significantly more energy than women. This is not due to the fact that women are smaller on average and require less "building material." Even when small men are taken into account, we come to the same conclusion: women live using less energy—and they live longer. This refers back to the "life candle" that burns out more quickly the brighter it glows.

The connection between increasing energy consumption and fast physical wear and tear seems to make sense. Other observations have likewise shown that decreased activity is linked with a longer life. Animals with shorter life spans have significantly faster pulses than those with long life spans. With every heartbeat, a second goes by on the clock of life. Whoever has the fewest heartbeats within a specific range of time lives the longest. This can be inferred from the following simplified overview:

|  | Heartbeats/Minute | Maximum Age |
| --- | --- | --- |
| mouse | 650 beats | 4 years |
| a bat in flight | 850 beats | 24 years |
| (a bat in hibernation | 200 beats) |  |
| cat | 240 beats | 35 years |
| elephant | 46 beats | 70 years |
| adult human | 80 beats | 108 years |
| whale | 15 beats | 100 years |
| turtle | 20 beats | 130 years |

Only mammals, animals that produce their own body heat, are shown on this table. Other animals, especially reptiles and fish, whose body fluids have the same outside temperature, do not show any correlation between pulse and age. This sort of data can only point toward the reasons for aging, but they cannot explain them entirely. For example, a squid with a heartbeat of thirty-five beats per minute does not live for nine years, as the table above would suggest, but only for two or three. Similarly, snails seldom grow older than twenty years, although their hearts do not beat more than fifty times per minute.

The relationship between a faster pulse and an earlier death can be applied only indirectly to humans. This is demonstrated by a research project that began seventy-five years ago and ended only recently. Beginning in the 1920s, the psychologist Lewis Terman launched a study of the lives of 1,528 Californian students who had high intelligence quotients. At the start of 1995, the last team of this research project evaluated the personality tests and the death certificates of all those who had died (over half). None of those involved in the test had died of hunger or other environmental conditions that would have distorted the results.

Two factors that crystallized from this study showed a significant effect on life span. Divorced parents, described as a factor for "social instability," cost those affected four years of their lives in comparison with the average age of death. And "good" people lived significantly longer than average. Earlier research, by Harvard psychologist Howard Friedman, noted, "Children, boys in particular, who demonstrate good sense, conscientiousness, honesty and humility in their personality tests live much longer. What we did not expect, however, was that happiness went along with a shorter life span." The journal *Science* explained this by observing that "jovial and open children are often at the same time impetuous, egocentric, and arrogant, which is related to drinking, smoking, and higher risk-taking." Behavior that shortens life.

There are countless other factors that indicate that lower energy consumption can lengthen life. If flies are held in small containers where they cannot fly, their life span lengthens significantly. They eat less and consume less oxygen. Mice who are given only two-thirds the normal amount of nourishment live about 30 percent longer than mice who are allowed to eat as much as they want. The reason for the longer life span has to do mostly with reduced oxygen intake. Oxygen is not only the life-giving component of the air we breathe; it is also, chemically speaking, a very aggressive gas.

A human being's oxygen intake by means of energy consumption can be purposely reduced in two ways: through sleep and coldness. These two conditions can allow for the prolongation of life.

---

### Free Radicals

Oxygen atoms generally exist in pairs. Sometimes, however, an individual atom separates itself from the pair and becomes a highly reactive "free radical." Such free radicals can originate in the cells of a body through chemicals or through energy sources, as in X-rays or ultraviolet light. They attack other molecules in the body, among them, DNA and unsaturated fatty acids. Although the body can repair much of the damage,[10] many larger and smaller defects develop over the course of time that subsequently reduce bodily resilience. It seems reasonable to assume that free radicals constitute one of the main causes of aging.

Antioxidants block the actions of free oxygen radicals. Included among them are vitamins found in cruciferous vegetables, like broccoli, cauliflower, and kale, and in root vegetables like carrots and butternut squash. In fact, the life span of animals fed antioxidants increases up to 50 percent. But if influences such as smoking are introduced, some antioxidants can also shorten the life span. No one really knows why this is the case. But it is clear that it doesn't pay to live an unhealthy lifestyle and, at the same time, try to ameliorate those effects with the aid of antioxidants in powder or pills.

---

Professor Heltmeier of Marburg was able to demonstrate through direct measurements made on marmots that an energy level of only 80 microwatts per gram of body weight is necessary to keep them alive, but barely. According to this measurement, ten people could be kept alive in a cold slumber with the same amount of energy necessary to power an average 60-watt lightbulb. This would not last for long, however. A normal heartbeat and an active brain require at least three times this amount. In addition, humans cannot convert electrical energy into metabolic energy. Therefore, the human body will feed off its own energy reserves—and this can occur slowly, while the body sleeps, for example.

During hibernation, marmots require only 5–15 percent of the energy they would consume otherwise when they are active. A comparable long period of sleep with short waking phases for the consumption of food could lengthen human lives (purely mathematically) ten times over! Even if these people were to spend half of each year in artificial sleep, they could become 350 years old, on average.[11]

On the other hand, the marmots sleep so deeply that Vitus Dröscher describes it as an "uncanny or frightening" state. Dröscher reports the following study, apparently peculiar in its details: researchers dug up a marmot that was six to nine feet underground. He writes,

> It felt stiff and ice cold and had rolled itself up into a little ball. The scientists tried to awaken it by poking its body with a needle, by cutting its tail with a knife and by playing a trombone near its ear so loudly you would have thought it was Judgment Day. Finally, they threw it like a rock off a cliff. The marmot never woke up.
>
> The winter sleep of hibernation is a kind of state of death, or suspended animation. And this is something that scares us to the core.[12]

The fact that the marmots' gruesomely deep slumber at a body temperature of only 40°F saves energy is illuminating. Whoever lies in an almost deathlike state of sleep six to nine months of the year and dozes a lot the rest of the time is likely to live a long life.

One wonders, then, about the fact that efforts like climbing stairs can lengthen life as well. More energy and oxygen are consumed in the process, which in turn leads to an increased level of free radicals in the body. All the same, scientists at Johns Hopkins University in Baltimore determined that every stair taken lengthens the stair climber's life by four seconds. After forty years in the workforce (up the stairs in the morning, down the stairs in the evening), with an apartment on the second floor, a "life-credit" or "time-reserve" of at least three days is created. With a bit of effort and more stair climbing, up to a week of additional time could be added to the reserve.

The reason: during athletic activity (including stair climbing), the body not only consumes more oxygen, it also creates more antioxidants. Perhaps it is these antioxidants that make stair climbing so beneficial to our health.

Professor Roland Prinzinger of the University of Frankfurt also subscribes to the theory of a "life candle" that burns down all the more quickly the brighter it shines. Since the late 1970s, Prinzinger has been studying how much oxygen very young birds consume during growth. He discovered that the only factor that determined how long the embryo remained in the egg "before it cracked open and was born as a chick" was the amount of oxygen consumed. Animals require oxygen for one purpose only: to change the components of their nourishment in such a way that it can be metabolized and used. Without oxygen, we could eat as much as we wanted—but we would collapse feebly because our bodies could not transform nutrients and oxygen into energy, thus requiring the brain to shut down. The more oxygen an animal takes in, the more energy is consumed.

Prinzinger, a metabolism specialist, compared the constant level of oxygen intake among the bird eggs with the observation that all bird embryos require the same amount of nutrients before they are ready to hatch. This is the case for all species of birds. From four-inch wrens to three-foot swans (*Cygnopis cygnoides*, found in east Asia) to the enormous albatross, all bird embryos ultimately need the same amount of energy to hatch. While the wren hatches after only two weeks, an unhatched albatross gives itself almost three months' time. The still tiny albatross consumes far less energy per minute in the egg than a nimble wren. Do albatross fetuses live more slowly?

Further experiment results and measurements convinced Prinzinger that for the most part the life span of *adult* animals and people is dependent on amounts of converted energy. In addition to the birds with which he has worked extensively, he cites observations from daily life. "As prowlers, cats are famous for their love of long naps," the biologist and chemist writes in his book, *The Secret of Aging*. "They live up to twenty-five years, significantly longer than highly active hunting dogs. Such dogs generally live fifteen to eighteen years. Dog species that live in rather cold regions that have very high metabolic rates tend to have especially short life spans. Typical examples are relatives of the sled dog, which generally do not grow older than ten or fifteen."

The notion that the consumption of oxygen is an indicator of energy consumption (that is, for the metabolic consumption of food for energy) and directly linked with life expectancy is impressively simple. However, Prinzinger's colleagues in the medical research of aging remain apprehensive about his apparently indisputable proof. Above all, they have difficulty believing that the multifaceted mystery of aging could be solved by such a simple explanation. This seems to annoy Prinzinger: "This shocks me to some extent. Why does everything have to be complex? Are not the most sophisticated solutions of nature at times stunningly simple?"[13] Prinzinger can even visualize to himself the biological clock with which the body measures the amount of energy it consumes. The breakthrough of metabolism theory in scientific circles is a highly sticky matter. Perhaps this has to do with the fact that we are just learning how intricate and detailed the system by which life develops from egg to adult animal is. Researchers cannot or do not want to believe that the consumption of energy is the sole determinant of how long the balancing act of life will last. Nonetheless, metabolism theory is hardly new or untested. Quite the contrary. As far back as 1908, Professor Max Rubner of Berlin had already published the first scientific treatise on the subject. He figured that adult animals almost always consume the same amount of energy per kilogram of body weight during the course of their lives. Since the processes of bodily wear and tear are of minimal importance in metabolism theory, this construct seems conclusive. Perhaps it is precisely the splendid elegance of the image of a self-consuming candle that made Prinzinger's colleagues hesitate to accept this explanation of aging.

## DO SPORTS STRENGTHEN VITALITY?

Gerhard Uhlenbruck, professor emeritus of immunology at the University of Cologne, is truly one of a kind. He does not hold much with formalities, which in German professors is quite unusual, and he has a terrific sense of humor, documented, among other places, as quotes on calendars that are available for sale in every bookstore in Germany. His workroom, with a view over a four-hundred-year-old cemetery, is brimming with books, and includes a lamp in a style from the late 1950s. In his eventful life, this vibrant, unconventional scientist and physician has successfully treated many victims of cancer—without medication. His preferred form of therapy: endurance sports.

The relationship between sports, the immune system, and the process of aging is based on the aforementioned observation that age measured in years of life is often higher than a person's physical and biological age—exactly in agreement with the adage "You're as young as you feel!" Uhlenbruck takes it even further. He maintains that the feeling of youth is not an illusion but is actually reflected in the immune system. The entire resistance mechanism of the body, according to Uhlenbruck, is "exactly as old as the spirit (or rather the brain) that gives it life."

Our immune system does in fact age, but we can slow down the aging process. As we age, our immune system does not slow its functioning the way other organs do, rather, its reactive powers change in strength.

Some defense mechanisms become stronger with age, others grind to a halt.

Doing sports regularly can break the aging process of the body's defense mechanisms: a sport trains not only muscles but the components of the immune system as well. How is this possible?

Every strenuous task places the body under stress. The body must respond to this with protective measures (one of which is fatigue). Those who participate in sports are relatively good at measuring when their training capacity reaches its maximum, or, in other words, when a break is necessary. Intentional stress situations, like athletics, are called "good stress" and differ greatly from dangerous, detrimental stress. (Unwanted, extreme stress situations, as they occur in violent accidents, for example, do not have resulting positive effects on the immune system.) Good stress causes microscopic breakdown within the body's muscle cells, and the body responds as if it detected a series of smaller injuries. When the body recuperates, it rebuilds new tissue and allows the body's muscles to grow stronger. In addition, immune cells may respond especially when "called" by injury—normally there are only a handful of "guard cells" that set off an alarm—the intentionally caused smaller group of infections requires constant attention. More and better equipped immune cells circulate continuously in the blood. Cells that have the *potential* to function are changed into cells that function *actively*. The immune system is trained to recognize possible injuries immediately and responds more quickly than an untrained system when a serious infection arises.

Immune system training grows increasingly important as we age because we are exposed to more and more environmental hazards. Once professional and private goals have been reached in life, if trust in one's own mental and physical abilities wanes, the motivation to try new things can feel daunting. Even sexual activity is often put aside. Chronic illnesses can weaken the immune system, since the body requires the totality of its resources all the time—there is neither the energy nor the desire other than to try to be well. Many of these processes are interrelated, thus making it difficult to combat them individually (with medication, for example). An aged immune system cannot respond effectively to all these varied influences. As a result, many so-called autoimmunities, composed of immune system proteins that attack parts of the body, are found in the blood of older people.

It is obviously more beneficial to keep the immune system younger. Professor Uhlenbruck helps critically ill people get back on their feet (indeed, running/walking is a suitable endurance sport) and helps people protect themselves against aging. The most important principle: "Regular endurance sports that one enjoys—at least three times a week for forty-five minutes. Competitive sports are less appropriate as we age: too much of a good thing is not so good. Nonetheless, the private experience of success should not be summarily rejected, even if it does not contribute directly to the strengthening of the immune system."

Athletics promote other life-lengthening processes in the body. One example is good sleep. Sports are useful for people with sleeping disorders, since they combat the lack of a certain substance normally produced by the immune system called interleukin 1B (beta), which is common among insomniacs. At the same time, more blood is circulated through the brain, which receives more oxygen. One can think better and in general gain a better sense of well-being.

The only real danger to sports is their potentially addictive quality. When sports are no longer a daily given but become an end in themselves, it does not take long before problems in interpersonal relationships arise. More than one love relationship has been destroyed by greatly exaggerated athletic goals. Professor Uhlenbruck writes: "Physical well-being should be the basis for enjoying other valuable things in life: intellectual and artistic interests, travels, love, and sexuality. Taking care of emotional needs through interpersonal relationships is critical, and above all, whoever has a sense of purpose in life does not lose himself or give himself up!"

People in industrial nations are so used to their creature comforts that they have still not found a comfortable substitute for regular endurance athletics as an integral part of their lifestyle. Much of what is achieved through sports can be achieved by pills—at least in part. In the United States, there is an entire branch of medicine devoted to "geriatric doping," that is, pepping elderly people up with drugs. With this method, athletic ability is not increased but made superfluous instead. The pills these people ingest are nothing other than growth hormones. In old age, the body ordinarily does not dispose of these substances; it has in most cases fulfilled the function of reproduction and does not need to remain young. Drugs can replace growth hormones and prevent premature aging. It is clear, however, that taking several drugs or medications as a fitness replacement is exorbitantly expensive and, due to its limited effectiveness, is simply nonsensical.


## THE "FRENCH PARADOX"

Basic reason proves that sports increase our strength. What is less easy to comprehend is that many activities ordinarily categorized as hazardous to one's health can also have a life-lengthening effect.

Researchers Brad Rodu and Philip Cole, working at the University of Alabama, Birmingham, reported in the journal *Nature* that consumers of "smoke-free" tobacco (chewing tobacco), live seven or eight years longer than tobacco smokers. This statistic, which at first glance does not seem particularly remarkable, is not accurate, according to Don-John Summerlin from Indiana, an expert on diseases of the mouth and jaw. Rodu and Cole overlooked the fact that, among consumers of smoke-free tobacco, cancer of the tongue and jaw are the most common forms of cancer.[14] This points to the fact that tobacco, with or without smoke, always attacks neighboring tissue. But the statistic is still puzzling. Perhaps tobacco smokers really die eight years earlier than those who chew or snort it, but, more likely, all that can be interpreted is that chewing tobacco shortens life less than smoking it.

A similar phenomenon can be seen with alcohol consumption. In twenty-seven countries around the world, the rate of people with heart diseases rises in accordance with the consumption of beer. Only in countries where red wine is preferred is the statistic lower. This phenomenon is known as the "French paradox," since, on the one hand, the French consume the most alcohol (apart from the Russians, about whom there are not equally reliable statistics), and, on the other hand, have the lowest rate of heart disease mortality—other than Japan. Could Psalm 104:15 really be true, according to which "wine gladdens the heart of man"?

It is more probable that a chemical substance common in red wine most generally and in wines from Bordeaux in particular actually protects the heart. It is found only in red wines because it is the only sort of wine produced with the skins of the grape; the substance is found primarily in the skin and protects the grapes from fungal infections. This wonder drug was discovered only a few years ago by biochemists Evan Siemann at Rice University and Learoy Creasy at Cornell University. Serge Renaud, director of public health in Lyon, advises that "for the prevention of heart disease, there is no more effective medicine than modest alcohol consumption." This means approximately one to two glasses of wine or two to three bottles of beer per day, equivalent to 20 to 30 grams of pure alcohol.

Why then does Monsieur Renaud speak of the medicinal advantages of alcohol in general and not of red wine specifically? Because, it would seem that in his view, as well as in the view of the World Health Organization (WHO), beer and any other alcoholic drink can protect against heart disease as well. The WHO claims that a quart of beer per day can protect people from heart disease and heart attacks; for women, two cups suffice. The study expressly warns against greater alcohol consumption, however. The appropriate total amount of alcohol consumed daily plays a critical role. This begs the question, then, to what extent the chemical substance by the name of resveratrole found in red wines only is really responsible for heart protection.

This is possibly a case of an unexplained cause and effect relationship, the old chicken-egg dilemma. Perhaps death from heart failure in countries where a great deal of beer is consumed is actually the result of high-fat diets or stress from traffic accidents or any other cause. In any case, the ape that played Cheetah in the Tarzan films holds the record for aging regardless of high alcohol consumption. "The odd chimp, who is going on 65, is kept fit by animal trainer Dan Westfall," reported the Rhine-area daily paper *Express* in 1996. It would seem that the regular regimen of physical training conducted by Mr. Westfall was necessary, considering the "eleven beers a day and a few glasses of schnapps to boot" that the chimp drinks. All the same, his drinking habits most definitely have adverse effects on his body. "The chimpanzee has a beer-belly and weights about 160 pounds," concludes the newspaper's brief report. It remains especially remarkable, however, that Cheetah has reached the age of sixty-five—even under the best of conditions, chimps in zoos do not ordinarily live beyond fifty.

### Luxuries for the Elderly

As German researcher Hans Franke reports, "Elderly people often have the need for small amounts of alcohol. According to a survey of 548 aged people, more than 65 percent drink small amounts of alcohol—50 percent of the women and 75 percent of the men. Most men who were asked drink 1 to 2 glasses of wine per day, and red wine seems to be preferred, perhaps given the old adage by cartoonist (creator of the famous German cartoon, *Max and Moritz*) Wilhelm Busch: 'Red wine for old guys is among gifts most divine.' The rest of the elderly studied indulge in a bottle of beer or a glass of Schnapps daily. Very elderly women drink a glass of wine or a bit of cognac each day. Real drinkers are relatively unknown among this group of elderly people."[15]

In Germany, old people's preference for coffee and tea is also prevalent. This does not stem from any need to compensate for "bad times" (i.e., wartime) in which these beverages were seen as rare, expensive luxuries: a study from 1930 reported that 75 percent of those asked said they drank up to seven cups of weak tea or coffee daily. Hans Franke notes instead, "Elderly people seem to fight daily fatigue with such habits."

In addition, not a single person surveyed in this group ever took medication for the purpose of lengthening their life—neither the so-called natural elixirs, like ginseng, vitamins, and various others intended to strengthen the heart, circulation, and endurance, nor prescription drugs that contain chemicals, live cells, sex hormones, and more.

## SMALL SECRETS OF PEOPLE A CENTURY OLD

Lady Barbara Cartland, who died in her ninety-third year and was the ex-stepmother of Lady Diana, swore by a combination of calcium and vitamin C. During the course of her life, she wrote hundreds of romance novels, each and every one of which achieved dream sales of more than one million copies. In 1994, she astonished her publishers with her sixth autobiography, at the same time giving away the secret to her fitness: she surrounded herself with and dressed exclusively in the color pink. This is astonishing if for no other reason than the fact that a study of Alzheimer patients in France done over a long period of time demonstrated that mental activity protects the brain—not color perception.

The French researchers discovered that people who completed high school and went on to intellectually challenging careers died far less frequently of Alzheimer's than people who completed high school and afterward (according to the research team) performed less intellectually and mentally challenging work. Professor Meier-Ruge of the Institute for Pathology at the University of Basel also confirms these findings: twice as many people who live in small towns and villages die of Alzheimer's than city dwellers. This assumes, of course, that life in small towns and villages is less mentally stimulating than the hectic pace of life in big cities.[16]

While only three people lived beyond the age of one hundred in 1938 in Germany, today there are over five thousand people over one hundred. There are twice as many in the Netherlands as in Germany. The legend that old people from the Andes and Caucasus mountain regions, whose secrets of longevity include the regular consumption of kefir (a beverage made from fermented cow's milk) and melted butter, is often promulgated by the press but never proven. Neither has any of these people ever been found.

---

### Kefir

The use of kefir as an anti-aging substance began around the end of the nineteenth century with Russian cell biologist Elias Metshnikov, who believed that "the acids of sour milk drive away poisonous, wild bacilli from the intestines." Metshnikov had apparently heard about extremely old people in Bulgaria and became convinced of the value of "Bulgarian milk-bacilli."

Metshnikov's theory became very influential, although it remained unproven throughout his life. He wrote many books on the subject and was even compared with Charles Darwin because of this discovery. In 1909, George Herschell published the book *Sour Milk and Pure Cultures from Milk Acid Bacteria for the Treatment of Disease*. Two years later, Loudon Douglas's book, *The Bacillus of Long Life*, followed—both written in the wake of Metshnikov's research. It is therefore no surprise that the erroneous faith in the power of kefir has persisted to the present.[17]

---

While the ancient drinkers of kefir remain elusive, the life and death of the second oldest person in the world has indeed been documented.

Shigichio Izumi of Japan died in 1986 at the age of 120 years and 237 days. For this reason, he was honored with the title of "national monument" in his homeland. Shigichio's tips for an active old age are not known. We know only that he gave up smoking at the age of 116 and drank a Japanese liquor known as Shochu daily until the very end of his life. However, a comprehensive analysis by Vaino Kannisto, a distinguished researcher at Max Planck Institute, Rostock, has shown that an age of 120 must be seen as an absolute exception. Kannisto studied the age of death of more than fifty thousand people who lived to be over 100 in fourteen industrialized countries. The highest age reached by most is around 108 years. The human body can neither withstand destructive environmental influences, nor the ticking of its own internal clock.

Particularly unusual is the fact that Shigichio Izumi was a male. Women generally outlive their husbands by five to eight years. This is most likely due to the heart-protective effects of the female sex hormone estrogen. Up until a few years ago, women's higher age average was also explained by lower alcohol and nicotine consumption. The Metropolitan Life Insurance Company predicts that men and women will have more equal life spans in the future. The reason for this, the company explains, is that "more and more women are assuming a higher-risk, typically male lifestyle" and thus die just as early as men.

Two facts may speak against Metropolitan Life Insurance, however. First, alcohol consumption, when moderate, can have a life-lengthening effect (see above). Second, an early death seems to be caused less by "male" behavior than by small, gender-specific differences in DNA. There are grounds for this assumption, since women have two X chromosomes (DNA that has the form of the letter $X$ in humans), while men have one X and one Y chromosome. (In contrast, mice chromosomes are both Y-shaped.) And an observation from the animal kingdom supports this: male birds, whose X/Y DNA is the same as females', live just as long as their female partners, although they experience their lives in an equivalent "risk-taking, masculine" way.

8 Jeanne Louise Calment died on August 4, 1997, at the age of 122. The photograph shows her on her 122d birthday on February 12, 1997. (Photo: AP)

### The Ultra-Aged

Not all the following reports about people over the age of 100 from various sources would necessarily be proven true. Often covered in sensational tabloid newspapers, they are nevertheless a testament to the dream of a very long, perhaps even eternal life.

- Bushtai Brezenian allegedly reached the age of 138 in 1991. At the time of his death, he was living with his gay partner, Giorgi Salmessi, who supposedly reached the age of 130 in the Caucasian Republic. The couple apparently fell in love in 1881 (*Independent,* December 24, 1991)
- West Africa's ostensibly oldest and most feared witch doctor (called a *baoura*) was Bawa Daouda. He allegedly died in January of 1994 in a region of Nigeria known as Bagari, the place where he was born in 1868. It is said that he still had all his original teeth at the time of his death; his youngest child was 115 at the time. Since he had tremendous magical powers, he attended his patients from behind a curtain so that they would not be overwhelmed by his aura. He could influence birds and insects to attack enemy armies and called Allah a comic actor. (AFP, January 23, 1994)

*(continued)*

- Jeanne Louise Calment from Arles in southern France died on August 4, 1997, at the age of 122 (figure 8). When she was fourteen, she met the painter Vincent Van Gogh, who bought canvas at her father's store. Mrs. Calment found the artist "moody, very ugly, and unfriendly," and he also "stank of alcohol. We called him 'Dingo.' " Mrs. Calment's daughter died at the age of 63—about 60 years ago. Jeanne Calment's advice: "If you want to live a long life, make sure you have a lot of fun. I know that I am going to die laughing." (*Guardian*, March 3, 1993)
- The piano teacher Carrie White died at the age of 116 years and 80 days in a hospital in Palatka, Florida. In 1909, she was sent to Florida State Hospital because of a typhus-induced psychosis, and lived there for 75 years until she was moved to Palatka. In 1988, the *Guiness Book of World Records* named her the "official oldest person in the world." (AP, February 15, 1991)
- In Ireland's West Glamorgan, mine worker John Evans died on June 10, 1990, at the age of 112. At the age of 73, he had been forced into retirement because of work regulations. He worked in his own vegetable garden until he was 95, at the age of 108—considered the oldest patient ever to have lived—he was given a pacemaker, and at 110 he visited London for the first time. His prescription for a long life: "No tabacco, no alcohol, no cursing, and no games." Each morning he ate a bowl of bran and drank a cup of hot water with a bit of honey dissolved in it. (*Daily Telegraph*, June 11, 1990)

## LIFE-LENGTHENING NUTRITION

The basic prerequisite for a long life is optimal physical fitness. That's why it is necessary to maximize one's nutrient intake as much as possible. This idea was formulated by Giovanni Boccaccio in the fourteenth century thus: "We die younger from the damned vice of gluttony than nature requires. Many people die too early, validating the saying from the wise masters of medicine—that food kills more people than the sword." Of course, a well-nourished body, on the other hand, works better than one lacking sufficient fuel.

Certain kinds of food, aside from over- or undereating, are more valuable than others. Vitamins, considered particularly "healthy," play an important role. In order to understand how different nutrients can affect the extension of life, it is useful to get to know the three main components of the food we eat.

An adult eats approximately half a ton of food yearly, about seven times his or her bodyweight, without gaining weight. Most of the nutrients derived from food intake are "burned" by the body—the process is chemically similar to the burning of wood, but it proceeds much more slowly.

The body tends to store fat because it weighs relatively little compared to muscles, bones, and other dense tissues. The required fat intake for adults is around 50 to 60 grams per day, something that vegetarians (especially vegans), people who eat mostly plant products, dispute. Experiments by Yale researchers Thomas Osborne and Lafayette Mendel early in the twentieth century seem to indicate that fat can be replaced by other nutrients–rats survived without any fat intake for up to 277 days. People who physically exert themselves often need to eat foods with a high fat content to maintain their performance power. This could mean that fat, although high in energy, is not needed on a continual basis. There is another reason we cannot entirely abstain from fat: only with its help can our bodies absorb fat-soluble vitamins such as vitamin A. An adult requires some fat, and about 50 to 60 grams of protein per day. Proteins are not burned directly but are usually used as "building blocks" for various body parts, especially muscles and internal organs. To some extent, proteins produce the other two primary components necessary for a healthy metabolism: carbohydrates and fats.

When the body digests, the body's temperature increases, just as it does with sunburn or malaria. Eskimos take advantage of this by eating a lot of protein in the evening, thus producing greater body heat. Increasing the metabolism in this way, however, can also lead to weight loss.

Potatoes, starches, and sugar are our providers of carbohydrates. Sugar plays a subordinate role in metabolism and is consumed in greater quantities than necessary in modern society. Together with fats, carbohydrates are our greatest source of fuel. Our bodies store no more than 1.5–2.5 pounds of these nutrients.

Vitamins do not provide the body with fuel. With the exception of vitamin D and some forms of vitamin B, the former of which can be synthesized from exposure to the sun and the latter of which can be synthesized by some intestinal bacteria, our bodies cannot produce vitamins themselves. We are dependent on getting them from external sources. If we get too few vitamins, symptoms of deficiency can occur. On the other hand, taking too many vitamins can cause illness as well.

B vitamins work as helper molecules for the body's many different metabolic pathways. They are not exhausted, or oxidized, in the process. In this way, only a few vitamin molecules are needed, each of which can be used again and again.

Vitamin E is interesting, since it is touted so often as an invigorator, capable of increasing levels of fertility. Whether the additional intake of vitamin E has any effect at all on humans has not been proven, however, since the experiments done in this field have primarily been carried out on animals, whose metabolisms differ markedly from humans.

Most vitamin E experiments have been conducted on rats. When researchers withheld vitamin E from the rats, they observed several results. For example, rats lost total fertility, and existing pregnancies led to miscarriages. These results occurred in a controlled laboratory, and it should be noted that such extreme vitamin E deficiencies do not ordinarily occur in humans. A further discovery, which may or may not carry over to humans, was that vitamin E protected the surface of rats' red blood cells, called plasmalemma, lengthening the life span of those cells.

The most famous of all is vitamin C. Since its absence among sailors resulted in much dreaded scurvy, its chemical name derives from "anti-scurvy acid," or ascorbic acid. Vitamin C is water-soluble, which means that when we boil vegetables with vitamin C or potatoes, it goes partly into the cooking water.

The human body needs about 70 milligrams of vitamin C per day. In some places in the world, it is harder to get that necessary amount because the availability of those fruits and vegetables high in vitamin C are less plentiful during cold seasons.

Nonetheless, there is little need to supplement a normal diet with any additional vitamins, assuming one is healthy. The body cannot absorb the enormous amounts of vitamins that are contained in tablets and multivitamin drinks; the vitamin dosage in some of these products is sometimes more than one hundred times the recommended daily allowance. The excess vitamins are excreted from the body.

What is most often forgotten in the discussion about the "miracle vitamin" C is human biology. Without a catalyst, vitamin C cannot be absorbed into the bloodstream; it cannot accomplish anything—it cannot cure scurvy. Two of the best-known catalysts that allow for absorption of vitamin C are calcium and vitamin E. Vitamin C works best when it works in conjunction with either. While many fruits and vegetables naturally contain vitamin C together with vitamin E or calcium (think of leafy green vegetables), most often these vitamins do not combine in artificially extracted vitamins sold in pill form. Why not? The extraction of vitamins and the process of recombining them again for mass manufacture and consumption is simply too expensive. When you purchase vitamin pills, the vitamins contained therein are separated from one another and work differently in the body than do fresh fruits and vegetables. It can be argued that the ingestion of expensive vitamin pills results in little more than expensive human waste flushed away into modern sewage systems.

Consider: today 100 grams of vitamin C can be bought for about $3, but some manufacturers charge up to $30 for 60 small "Vitamin B Complex + C" tablets. Back in 1930, vitamin C was more expensive than gold, while other vitamins were all but ignored. In 1936, only 6.5 pounds of vitamin B1 were produced in all of North America—for a total price of $8.

Our obsession today with exaggerated intakes of vitamins via tablets, capsules, and drinks for the most part stems from those days when so many at sea died miserably of scurvy. The Portuguese explorer Vasco da Gama lost half his sailors to the disease during his voyage around the Cape of Good Hope in 1497. In the first half of the nineteenth century, English explorer John Franklin's Arctic expedition also failed tragically; in spite of excellent provisions designed to last for several years, the entire crew apparently died from a vitamin C deficiency. Trapped for many years in pack ice, these seamen had finally attempted to make their way to safer shores, but never reached them. When, years later, the remains of the dead men were discovered, large amounts of chocolate and other vitamin-free foodstuffs were found with them. Only fresh fish could have saved these men; they were eaten by fish instead.

## LINUS PAULING'S LOVE AFFAIR WITH VITAMIN C

Nobel laureate Linus Pauling (figure 9) had a special passion up until the end of his life (he himself even spoke of "my love affair")—ascorbic acid.

9 Linus Pauling researched the healing and preventive powers of vitamin C throughout his life. He is photographed here in the mass spectrometer laboratory with Ewan Cameron, then medical director of the Linus Pauling Institute. (Photo: Pauling Archives/LPISM)

At his private research institute in California (now part of Oregon State University), he devoted countless books and articles to the subject of vitamin C, investing tremendous monetary resources on the employment of scientists whose work tested and proved the effects of vitamins. On May 24, 1992, he wrote, as reported in the *Philadelphia Inquirer:*

My book, *Vitamin C and Colds,* was published in 1970. Since this book stated that a daily intake of three grams of vitamin C can prevent colds on many fronts, I thought that doctors would be highly pleased that they would have more time to treat patients with more serious complaints. Instead I was branded a charlatan and a quack. It became all the worse when I published a book three years later together with Dr. Ewan Cameron that described the value of high doses of vitamin C in the treatment of cancer. I have wondered for twenty years why authorities in medicine have such a negative stance. Maybe it has to do with the fact that doctors are always careful not to prescribe high doses of any medication. Perhaps a medical reason explains their pessimism. Perhaps pharmaceutical companies have a particular stake in preventing the use of inexpensive medicines. I, however, am as convinced as I always was that large quantities of vitamin C as well as other vitamins can both prevent and cure heart complaints, cancer, and many other illnesses.

Pauling's views are based on myriad experiments and observations.[18] What is most interesting, however, is that the amount of vitamin C required by organisms such as cats, dogs, elephants—and, indeed, almost all animals—is produced by those animals themselves. Animals need vitamin C to build collagen, an essential protein. Collagen is a structural protein; it forms tiny fibers that are necessary in the body's construction and maintenance of blood vessels, skin, bones, and teeth.

Pauling did notice that animals produce a remarkable amount of ascorbic acid themselves. He converted the statistics from the studies of these animals to human needs, determining a daily recommended allowance. This recommendation indicated about 18 grams per day—three hundred times more than the official recommended daily allowance at the time in the United States of 60 milligrams. The idea of vastly increasing the daily intake of vitamin C did not originate with Pauling but with the biochemist Irwin Stone. He promised Pauling that even 3 grams of vitamin C a day would increase his life span by at least twenty-five years. Indeed, Pauling died at the ripe age of ninety-three.

---

### Linus Pauling's "Plan for Optimal Health"

In addition to taking vitamin C, Pauling also made other recommendations for good health and long life. These recommendations most certainly will have a positive effect on health and life span, beyond the heavy doses of vitamin intake:

1. Take 6–18 grams of vitamin C every day.
2. Take 400–1,600 units of vitamin E daily.
3. Take 1–2 vitamin B tablets daily.
4. Take 25,000 units of vitamin A or a 15-milligram beta-carotene tablet daily.
5. Take a vitamin-mineral tablet every day.                    *(continued)*

6. Cut your sugar consumption in half. Do not use any sugar in coffee or tea. Do not eat heavily sweetened foods, particularly desserts. Do not drink soft drinks or milkshakes.
7. Otherwise eat what you want—but exercise portion control. Eggs and meat are acceptable. Eat fruits and vegetables. Control your weight.
8. Drink plenty of water every day.
9. Stay active, get exercise, but don't overdo it.
10. Drink alcohol in moderation.
11. Do not smoke under any circumstances.
12. Avoid stress. Find a career you enjoy. Be happy with your family.

There are indicators that show the potential success of vitamin C therapy based on Pauling's many experiments and scientific investigations. The scientific *proof* of the effects, however, has always been debated. In many cases, people were cured inexplicably—without the use of modern medicine. Was it vitamin C? Today no one questions that the immune system is significantly influenced by more than vitamins or medicine. The immune system seems to respond to our own neurology, maybe even the psyche. Even banal illnesses like colds and coughs can be psychologically induced. Some people go without colds for years because they feel particularly good or when they are so busy with work that they simply believe that they can't afford to get sick. And so it goes: taking vitamin C with the total conviction of its health benefits can make for a kind of healing, similar to the placebo effect. In contrast, according to a study by American researcher Mark Levine, the body needs no more than 200 milligrams of vitamin C daily. This is easily obtained in Western industrialized countries, where it can be attained easily from fresh fruit. Only about three ounces, or 100 grams, of bell peppers, blackcurrants, or various kinds of cabbage contain up to 100 or more milligrams of vitamin C.

We know today that heavy doses of vitamin C can have negative health effects—among them, the development of kidney stones. Those advocates of vitamin C argue against this, however. All the knowledge in the world does nothing to temper the popular beliefs that are fed incessantly by advertisements and other media.

The same is true of the endurance sports program recommended by the Cologne physician Gerhard Uhlenbruck mentioned earlier. It leads not only to directly measurable changes in the body but also to a heightened awareness of one's own health. If one intends to reach a ripe old age, this is perhaps the most important principle. It is a well-known fact that very old people overestimate the status of their physical and mental well-being. That is to say, they feel better than they actually are. Since they are satisfied with their lives in general, they do not tend toward excessive modes of behavior (for example, overeating as compensation for lack of attention, care, or emotional interaction with others).

---

### Vitamins and Violence

As far back as the Victorian era, the English were very concerned with the care and feeding of prisoners—if a prisoner was well-behaved, he received more nutritious meals. This very simple and often very successful means of behavioral control exists today in different forms. U.S. studies in the last twenty-five years have shown that prison food in the United States often has too little vitamin C, vitamin B1, and zinc. At the same time, other studies have shown that a zinc deficiency in particular can lead to depression and aggressiveness. Bernard Gersch from the town of Ulverston in England tried to use this mechanism to lower the number of aggressive outbreaks in English prisons. In 1996, one hundred prison inmates in Aylesbury volunteered to participate in Gersch's experiment (he raised approximately $130,000 for the work). He gave fifty inmates a pill containing fatty acids, vitamins, and zinc, and the remaining fifty took a sugar pill (placebo). It is questionable whether this kind of diet supplement really lowered instances of aggression. In any event, some of the inmates claimed that they felt better—whether this stems from the positive effects of the zinc or from the fact that these prisoners' meals were prepared with greater care remains unclear.

---

That is precisely the same recommendation made by Hippocrates, the physician from Greek antiquity, and later by Goethe's physician Hufeland:

1. abstinence from all things that cuts life short and
2. moderation in all things.

The fact that almost all very old people are 5–10 percent underweight when compared with Broca's rule (height in centimeters minus 100 gives ideal weight in kilograms) seems to be supported by such advice. In the three regions of the earth apparently inhabited by the most aged, people eat only about 1,200–1,900 calories of food per day.

Flies and rats live longer with a lower-calorie diet as well: calorie intake reduced by two-fifths lengthens the life of laboratory rats by one-third. Age researcher Roy Walford is in the midst of an experiment on himself: he is testing whether a diet of fifteen hundred calories daily allows him to become older than the average American. If more people would participate in such an experiment, perhaps we would know more about the relationship between age and diet. But even if it is true that diet can slow the aging process: who would want to spend a lifetime on a severely restricted diet?

## MELATONIN AND OTHER PROFITABLE WILD CARD CURES

Linus Pauling is not the only respected scientist to "fall in love" with a chemical substance and maintain that passion throughout a lifetime. Some researchers have hailed melatonin as a miracle substance as well.

The existence of melatonin, a hormone praised as a wonder drug, has been known for a long time. It is reputed to help all modern ailments, such as high blood pressure, loss of sex drive, weak immunities, electromagnetic field-induced sicknesses, and cancer, of course. Melatonin is but another in the long line of candidates for the prize of overrated substances. It is true that the production of hormones in the pineal gland slows, starting at the early age of six; by the age of forty-five it is only half as strong as in childhood. But regenerative effects of the pineal gland appear to be influenced by hormones like melatonin and other questionable compounds. The question of how the regenerative effect *really* occurs remains open. When American physicians Walter Pierpaoli and William Regelson (with Carol Colman) promise in their book *The Melatonin Miracle,* "No one needs to grow old in old age," they hit the nail on the head: not everyone has to show signs of age with the passing of time. But this has as good as nothing to do with pills and much more to do with a person's balance in their mental and emotional life.

A lot of pill-popping apostles would do well to examine the original work of scientists who research substances like melatonin today. "[There are] up to this point no reports on the slowing of the graying of hair, wrinkle development, or the general retreat of aging signs or risks of cancer," write melatonin researchers Reimara Rössler, Peter Kloeden, and Otto Rössler of the University of Tübingen. "The span of observation time is too short and the number of experimental subjects studied too small to draw such conclusions."

In addition, the body develops a tolerance for melatonin. "A dose that once had an effect," the Tübingen researchers write, "no longer suffices after, say, six months."

Neither melatonin nor any other substance can effect any real lengthening of life span. Some substances contribute to increases in general well-being and slow down or at least camouflage physical decline to a very small extent. Here the ointments enriched with hormones—intended to tighten the skin and erase wrinkles—come to mind. Skin ointments and creams only help when a person practices a complete health regimen in an effort to maintain a youthful body. Melatonin should remain what it was before melatoninmania: a reliable protection against cataracts, jet lag, and sleep disorder—especially among older people.

Americans show even more enthusiasm for all sorts of chemical quick fixes than do Germans. British microbiologist Bernard Dixon summarized this in an observation he made in 1994: "Trace elements in tablet form are a hit in the U.S. 'Have you tried selenium yet, or are you still doing zinc?' was one remark coincidentally picked up by an elderly woman from Philadelphia, who ate her way element by element through the periodic table." That the U.S. "wave of wonder drugs" is surging more and more strongly in Germany is something greeted enthusiastically by the German health food industry. A free magazine called *Felice*, devoted solely to the sale of supposedly rejuvenating substances (the text and pictures are taken directly from the U.S. edition), makes the following promises: "Physical Energy in Capsule Form—Discovery Awarded Nobel Prize," and "New Knowledge Shows Plant Extracts Protect from Illness." In the magazine, capsules and pills with the following contents are advertised: coenzyme Q10, selenium, chromium, zinc, plant-extract pills with phytons, carotine, calcium, and borage seed oil. *Felice* fails to mention, of course, that all these things are acquired simply by eating a well-balanced diet. It is not mentioned that every plant on earth is composed of "phytons" (= plant parts) and that, in some cities, tap water has more than enough of the daily recommended dose of calcium.

The absurd consequences of the general youth craze have been well-documented. Professor Volkmar Sigusch, who is a professor studying sexuality at the University of Frankfurt, tells of critically ill seventy-year-old men who complain of their inability to maintain an erection.

## BIORHYTHM: THE METRONOME OF LIFE

The continual return of the seasons, a symbol of the course of human life, gives us the impression that there is a basic rhythm to nature by which everything repeats itself. Blossoms open and close, animals migrate and return. What gives the world its cues for its rhythmic cycles? Is the smallest rhythm of life subordinate to a larger, overriding one? Would this overriding rhythm explain the birth and passing of all living things?

Working at the turn of the nineteenth century, the Berlin physician Wilhelm Fliess posed such questions about life in the early seventeenth century. The conclusions he drew, based on his observations of nature, are interesting for two reasons. First, they show that our life is controlled by a basic rhythm. Second, they make clear that the relationship between cause and effect is not readily apparent.

Fliess sought to uncover the internal order by which life works. He understood that women ovulated every twenty-eight days, so he asked his female patients to keep precise records of their menstrual cycles. An analysis of the data showed that the so-called norm interval of twenty-eight days happened as good as never.[19] Since Fliess was obsessed with the idea that "a pulse runs through all of life," he did not give up. He calculated that the deviation from the number twenty-eight was related to another number: twenty-three.

Fliess tested countless articles in medical journals and, repeatedly, friends' personal reports that could be described with some relationship to the numbers twenty-three and twenty-eight. It worked. Every process in life could be reduced to these two numbers, beginning with periods of time between births to the time of deaths within families. The time a plant buds and the time it loses its blossoms, how often hermaphroditic people are born, when children's teeth push through their gums, even the lives of many generations of one family: Fliess put everything into a mathematical context.[20]

A short example demonstrates his method: Fliess wrote that the Austrian-born composer Franz Schubert "was extremely prolific on four days of the year 1815—on August 19 and 25 as well as on October 15 and 19. On these days, he composed his most beautiful songs. The amount of time between the days is as follows:

- August 19 through October 19 = 61 days = 2 X 28 + (28 - 23) and
- August 25 through October 15 = 51 days = 2 X 28 - (28 -23)."

Apparently, in the seventeenth century, only a basic arithmetical operation and two numbers were necessary to place particularly work-productive days in connection with each other.

Much more comprehensively, Fliess documented the life cycles of two amaryllis flowers he studied for eight years. One of the flowers was an offspring of the other—thus, the plants were closely related. Fliess noted the exact time its buds grew, blossomed, and fell off. At first glance, no relationship could be detected among his numerical data. Fliess highlighted the relationship in the following way:

- "The time between the bud growth from one year to the next is the same as the time between the blossoming in the first year and in the second year plus four times twenty-eight minus four times twenty-three."

It seemed as though illness and—particularly interesting—death were subject to the order of the numbers twenty-three and twenty-eight. Fliess cites another example in addition to the amaryllis flowers: his calculations also apply to infirmity and the moment of death.

- Out of two times (twenty-three plus twenty-eight) people who fall ill with St. Vitus's dance (a disease of the nervous system marked by uncontrollable muscle spasms), twenty-eight are men.
- "A Little Wolfgang could walk after twenty-four times twenty-eight plus (twenty-eight to the second power) minus two times

twenty-eight times twenty-three days. He lived twenty-four
times twenty-three plus (twenty-eight to the second power)
minus (two times twenty-three to the second power) days."

Fliess was even able to calculate the ages at which Goethe, Bismarck,
Kaiser Wilhelm, and Alexander von Humboldt (as well as many other
prominent people of that time) would die using the numbers twenty-
three and twenty-eight. How is this possible?

We find the first indicator of error in Fliess's arduous calculations
when we try to determine in advance the age a person will be when they
die: it is possible only to determine the date of death according to the
basic numbers twenty-three and twenty-eight in retrospect. This contra-
dicts the basic principle that future events or developments can be predicted on the basis of a "natural law." Hence, the numerical basis of biorhythms cannot have the status of a natural law.

A further error in Fliess's work consists in the fact that the numbers
that Fliess co-opted from other authors' work were not always accurate.
Contrary to the example just described, three times as many women as
men did *not* suffer from St. Vitus's dance. If Fliess was able to incorpo-
rate such false data into his system, it must be assumed that *all* data can
be included in a biorhythm in one way or another.

But what of the calculations based on accurate original data, such as
the budding time of certain plants? Are the connections among these
numbers artificial, or do they reflect an essential characteristic found in
nature?

The answer: even in "real" cases Fliess was victim of his own skewed
logic, which can be proven using the following calculation of probability.
Suppose I take numbers at random, be they from my observations of nature or produced more or less randomly (e.g., from a computer program). How high is the probability that one of these numbers is divisible
by either twenty-three or twenty-eight, which can be subsequently categorized in Fliess's biorhythmic cycles? We would find a qualified candi-
date out of approximately one of every twelve chosen numbers. By
studying certain numbers that belong together (cf. the example of the
Schubert songs), Fliess artificially raised the probability of finding numbers that fit into his rhythm scheme. It forms a perfect circular argument: numbers that belong together because "they belong together,"
were placed into one kind of grouping. Thus a certain rhythm is "discov-

ered" that was assumed from the start. Fliess "cooked" the numbers to match his starting assumption. Without the assumed rhythm, the numbers would not have united and would not have produced a specific rhythm.

The rhythms that are deduced from Fliess's numbers are artificially produced. No one simply took the time to reevaluate the several hundred pages of his work with numbers other than twenty-three and twenty-eight. However, in 1928, the Zurich physician Dr. Aebely demonstrated that many biorhythmic number games also function with another numerical pair, such as three and five. It is amazing nonetheless that, with grim determination, Fliess was able to construct a view of the world that seemed to be accurate down to the very last detail. Hans Schlieper, one of Fliess's acolytes, extended the work beyond his master. In 1929, he attributed even more value to the numbers twenty-three and twenty-eight. Incredibly enough, this made it possible not only to reconstruct people's lives mathematically but also to study such disparate experiences as the "regular" appearance of certain dreams and the growth and development of human beings. Observing the mosaic of an armadillo's skin, Schlieper noticed "structures that catch an expert's eye as unmistakably authentic: indeed! These were the number of years between me and all my siblings, 'translated' into spatial dimensions and depicted as a graph!" Schlieper made no secret of the fact that the wish itself gave birth to the idea. In one of his published writings, he stated, "I never doubted that measures of time, the year in particular, would make themselves visible on living things." In only the rarest of cases, the act of wishful thinking can prove successful, unscientific as it is. For example, the French researcher Louis Pasteur was determined to find a cure for rabies, and the early stages of his work were marked by brute force practices that finally won out—despite worldwide ridicule of his theories at the time. Nonetheless, Fliess's ideas for the scientific basis of time and mathematical measurements failed, since they were based on the quicksand of circular self-containment.

An elegant proof of this system's fallacies is offered unwittingly by the journal *Bild der Wissenschaft*, which awards an annual prize for the best mathematical calculation of the current year by using only the four numbers contained in that given year (e.g., for the year 1929, what mathematical combination of the individual digits 1, 2, 9, and 9 would yield the total 1929?). Readers connect the four digits through simple mathematical methods to produce the number of the given year. No one draws the conclusion that the four digits of any given year must have magical powers simply because they can be calculated in some form to produce the year in question. It is little more than a mathematical parlor game.

To sum up: using simple rules, someone who is mathematically inclined can find a way to connect almost any set of numbers and call it into a rhythmic series. It is irrelevant whether the numbers are "authentic," or "inauthentic," or whether they were chosen randomly.[21]

The obituary Schlieper wrote for Wilhelm Fliess reads as follows: "Wilhelm Fliess, who was born on October 24, 1858, died shortly before his seventieth birthday. The year of his death, like the years of his sons' births, were invested with scientific meaning:

I.

His son Robert born December 29, 1895

His son Conrad born December 29, 1899

The difference between the two births is 1,461 days, which equals 4 years plus 1 leap day

1,461 is 51 x 28, which is the same as 23 x 28, plus one 28 squared.

II.

His son Conrad born December 29, 1899

Wilhelm Fliess died October 13, 1928

The difference between the birth of his son and his own death is 10,515 days, which equals 32 years plus 8 leap days, minus 51 x 23.

Now 51 x 23 is 28 x 23 + $23^2$.

Fliess would have considered these last words about his life and death highly appropriate; in retrospect we have to smile when we think of the dogged determination with which Fliess and Schlieper proceeded with their work. The calculations are meaningless.

At first glance, Fliess's building blocks of biorhythms seemed to be built on strong mathematical foundations, but they dissolved finally into nothing. All the same, his teachings do contain some kernel of truth, for, of course, there are—other—biological rhythms. Before we look at these, let's consider a few more examples that demonstrate the great difficulty even educated scientists have drawing true conclusions from mathematical relationships.

## CAUSE AND EFFECT DISAPPEAR IN THE TANGLE
## OF NUMBERS

During the nineteenth and twentieth centuries, the belief that married men lived longer than bachelors was widespread. Even if numerical facts show that there is a relationship between the life of a bachelor and a shortened life span, this does not necessarily mean that the relationship is causal. For example, it could have been proposed that bachelors remained single because they were sickly, or unattractive, or because they led unhealthy lifestyles. The numbers indicate only *some kind* of relationship between the facts of "bachelorhood" and "earlier death" or between "being married" and "longer life." This does not necessarily mean that the cause and effect are real.

The only way we can protect ourselves from drawing false conclusions from statistics is to read each report very critically. In extreme cases, reason fails in the face of false assumptions. Take the case that in some parts of Germany more storks build more nests when more human children are born. Since we know that storks don't bring children, we know immediately that the relationship "nesting storks and children born," while indeed true, cannot be the cause of the rate of birth. Why is it, then, that there are more storks in regions where more children are born? The answer is simple: the larger a rural region is, the more inhabitants it has and the more children are born. The number of storks rises with the size of the rural area as well, since the opportunities for storks to nest rises proportionally to the available area they inhabit. The size of the rural region influences the number of children and the number of storks in equal proportions.

In other cases, it is not so easy to maintain a critical distance from statistics. One of the most passionate discussions I ever experienced among scientists revolved around the medication thalidomide. Children born to mothers who had taken this medication suffered severe deformities, primarily of the arms and legs. In the beginning, there was no *statistical* proof that the administration of thalidomide caused deformities. This was used as the key argument for the defense in the trial that ensued between the victims of thalidomide and the pharmaceutical company that produced it. Still, when a scientist for the defense was asked whether he would prescribe thalidomide to his wife during pregnancy, he gave the short and precise answer "No." Why?

In the case of thalidomide, it was at first impossible to mathematically prove the relationship between the drug and birth defects. However, this does not mean that no relationship in fact exists. The statistics *were not able* to prove anything at first in the case of thalidomide because there were too few birth defect cases—a purely mathematical problem that had nothing to do with biology or medicine. The conclusion that there was no connection between thalidomide and deformities, simply because the statistics don't prove it, is false. In this case, mathematics say nothing at all. They can't say anything, for statistics may not be applied as mathematical tools when the number of cases—here, thalidomide deformities—is too small. It would be like hammering a nail into a wall with a screwdriver. The conclusion "the nail can't be hammered in" is totally unreliable because the inappropriate tool was used.

---

### Bachelors and Fly Sperm

A report published in several newspapers in 1994 made the claim, "People who kiss a lot live on average five years longer, U.S. scientists have discovered. Kissing strengthens the immune system." The immune system isn't the only thing that benefits from kissing, however.

Research animals in Professor D. von Holst's Bayreuth laboratory, called Tupai, or treeshrews, tend to bond in pairs. These pairs appear to kiss frequently. If a pair is "coupled harmoniously," their pulses sink significantly. Harmonious pairs die much less frequently of heart attacks and other cardiovascular diseases than nonharmonious pairs. In addition, the partners' heart beats and breathing rhythms synchronize, something that can be observed among human couples as well.

If researchers use the activities of bond pairing, frequent displays of affection, and regular sexual intercourse as indicators of "togetherness," the discoveries are astounding. In January of 1994, the journal *Nature* reported extensively on the life expectancy of fruit flies. When male fruit flies deposit sperm into females, they also deposit a substance that kills other residual sperm in order to prevent successful fertilization of other males. At the same time, the substance has a life-shortening effect on the females—the longer and more frequently the females were inseminated, the earlier they died. *Nature* was careful not to apply these data to humans, however. "Semen is bad for you—but only if you are a female fruit fly."

## A RHYTHMIC TIME CAVE

All higher-order animals have their own internal clock. To adapt themselves to the times of day and night, animals use the duration of daylight, the length of night, the timing of the tides, or other environmental factors. Like a mechanical pocket watch that has to be adjusted periodically, animals need to adapt themselves to the actual rhythm by events in their surroundings. This adjustment is important because the time that an animal seeks food, finds a mate, or hibernates depends on the time of day and year. By continual readjustment of their internal clocks, animals can maintain correct daily rhythm, even in bad weather conditions, as when several days of heavy clouds obscure the exact time of dawn or nightfall.

Everyone knows that people have internal clocks as well, noticeable in the jet lag we experience after long flights across continents and oceans from either the east or west. Once we arrive at a new destination several time zones away from our homes, we remain wide awake or grow dead tired at the most improbable times: when our internal clocks are set for night, the sun is shining brightly.

Another indicator of the internal clock: some people are able to wake up at any exact time in the morning without an alarm clock (like the author of this book).

With the help of external signals, the body can adjust itself to a new daily rhythm; the internal clock is by no means inflexible. An inflexible biological clock would be absurd, since the adaptation to changed environmental conditions is a basic prerequisite for all life. Nonetheless, the bodily rhythm must provide a strong, if not entirely rigid, beat that can change and realign itself with the real length of the day. This regulating "beat" does, in fact, exist. It was discovered through experiments with people who willingly secluded themselves in a bunker for up to one year (among them many students studying for exams).

---

**The Biorhythm for a Good Day**

The results from behavioral, learning, and skills tests, together with blood samples, make it possible to draft a complete daily schedule that will allow a person

*(continued)*

to maximize the biorhythmically high points of the day and circumnavigate the low points.

| | |
|---|---|
| 7:00–9:00 A.M. | Sex hormones are at a high point—indulge. |
| 8:00–10:00 A.M. | Highest threshold for pain (make dentist appointments for the morning!) |
| 9:00–10:00 A.M. | Short-term memory works best. A good time to do a quick review before tests. |
| 9:00–12:00 P.M. | Analytical thinking abilities at full speed: the best time for problem-solving and strategy planning. |
| 10:00 A.M.-12:00 P.M. | Greatest degree of alertness, speaking ability optimal: the best time for negotiations, job interviews, and conferences. |
| 1:00–3:00 P.M. | The afternoon low. You sag, whether or not you ate lunch. A short nap helps you stay fit for the remainder of the day. |
| 3:00–4:00 P.M. | Alertness increases again, long-term memory reaches its highest capacity. A good time to study or to memorize a speech. |
| 3:00–5:00 P.M. | Mood reaches its high point of the day. |
| 4:00–6:00 P.M. | Skills are at a peak: piano playing, typing, arts and crafts |
| 6:00–9:00 P.M. | Everything slows down. Relaxation. Mental energy wanes. |
| 7:00–9:00 P.M. | Time for tasting and enjoying: the five senses are at their sharpest |
| 11:00 P.M.–1:00 A.M. | The late creativity high. Whoever is awake at this time can think, write, invent, and compose out of the stuff dreams are made of.[22] |

The initial assumption of the researchers maintained that people were "set to twenty-four hours." Only in 1970, once the bunker experiment began, did it become apparent that this was an error. The people in the bunkers did what they wanted and could ask for meals or supplies of games, paper, pens, etc. whenever they wanted. The researchers were interested most of all in when the subjects of their study turned the lights on and off. Since they had neither natural light nor clocks to act as cues, they heard no noises from the outside, and had no conversations with anyone outside the bunker (except to order food or supplies), a "day" was determined solely by the length of time between turning on the light and turning it off.

During the course of the experiment, most people kept their lamps on for longer and longer periods. Their days generally lasted twenty-five rather than twenty-four hours. They continually lagged behind in real time until they had "lost" several days, mathematically speaking. After the twenty-fifth day in the bunker, the test subjects started receiving daily papers from the day before, after the fiftieth day, the newspaper from two days earlier. "Only the twenty-four-hour rhythm of the Earth's rotation forces the rhythm of the cosmos on the twenty-five-hour man," explains the Göttingen scholar of experimental medicine Friedrich Cramer of his colleague Jürgen Aschoff's studies.[23]

In addition to twenty-five-hour-day keepers, there are apparently people with a shortened daily rhythm as well. Since their fundamental rhythm functions over a twenty-three-hour period, outside the bunker they tended to be early risers. The majority of the twenty-five-hour people tended to be late sleepers and moody in the morning.

The bunker experiments proved that people manage to level out their day and night rhythm during the course of the day. What place is left for Fliess's biorhythm? One would assume that there is no place at all. Amazingly enough, though, the old notion of biorhythm theory corresponds to a similar time indicator: the moon. The rotation of the moon around the Earth lasts twenty-eight days. Women ovulate within a similar time span—recall that this was the basis for the idea of the biorhythm. Could it be that the sun determines the length of day while the moon helps us to adjust the length of the month? The difficulty of proving such observations can be demonstrated by the following experiment.

In working and living community spaces (offices, apartments shared by women, prisons for women, etc.), it seems that women tend to menstruate at the same time. The fact that the women menstruated at different times before they entered a common communal environment begs the question of why this could happen at all. One research group tried to throw the menstrual rhythm off balance. Sweat from a menstruating test subject was applied to the bodies of female prisoners. Indeed, an imperceptible signal spread from the "pheromonal" substances in the women's sweat so that the rhythm of the other women shifted accordingly. The phases of the moon in and of themselves are not the only factors that determine the biorhythm, but they serve as a constant temporal frame of reference by which biological processes can be aligned.

It is assumed that the actual basic rhythm of human life is encoded in the DNA. Under many influences, this rhythm adapts itself to the length of days and other environmental conditions. In the end, we cannot dismiss or overcome our biological clock. Or can we? If we succeeded in genetically resetting our biological clock, couldn't we slow the inexorable aging process as well?

Genes (on DNA) do in fact determine aging. It is easy enough to see that there are families whose members tend to live longer lives. On average, members of such families die ten years later than people from the general population born in the same year. There are also families in which the male members, specifically, fathers, grandfathers, great-grandfathers, and so on, tend to live longer. This too is an indicator of an inherited longer life span, since under certain circumstances some genetic traits can be passed on exclusively by the paternal (or the maternal) line. According to what we know today, a long life is determined about two-thirds by genetics and one-third by environmental influences. Environmental factors are numerous and varied, including many mentioned already, as in marital status and eating habits.

It is possible to locate the areas within DNA that are responsible for purely genetically inherited long life. Perhaps one could then transfer these DNA components to other people, depending on what special features they possess. This process is known as somatic gene therapy. The question of what meaning this process could have in the future leads us to the third section of this book.

CHAPTER 3

# THE IMMORTALITY OF THE INDIVIDUAL:
# POSSIBILITIES AND (FOR TODAY) IMPOSSIBILITIES

---

*What does long life mean? The terror of being trapped in a human body whose*
*abilities diminish; the insomnia that is measured decade to decade, not with clock*
*hands of steel; the weight of oceans and pyramids, of old libraries and dynasties, not*
*knowing whether I am condemned to my own flesh, to my repulsive voice, to my*
*name, to a routine of memories, to the Spanish people with whom I cannot get along,*
*to nostalgia for the Latin language I did not master, to wanting to sink into death*
*and not being able to, to be and to endure.*

*JORGE LUIS BORGES*

## WHEN IS A PERSON DEAD?

Up until the 1960s, it was easy to determine death. If a person's heart or breathing stopped, the person was pronounced dead. Scores of people who would have been pronounced dead by these criteria can today move their legs, have erections, and, by reflex action, reach out and embrace their caregivers as their beds are being made. What has happened since the 1960s that has altered the determination of death so radically?

The invention of the heart-lung machine and the artificial respirator made it possible to maintain blood circulation and breathing in a person after the heart had, in fact, stopped; lack of a heartbeat alone no longer constituted death. In 1968, a commission of the president of the United

States established an entirely new criterion for determining death: brain death.

After the heart stops beating, it takes only about an hour before the energy reserves of the body's cells are completely depleted. The muscles stiffen, and rigor mortis begins. The body then decomposes at a rate dependent on the ambient temperature through the actions of fungi and bacteria as well as through the rapid toxic degeneration of the body's cells. The situation is quite different when someone is brain dead. The myriad parts and functions of the human brain vary in degree of conscious and unconscious importance. Deeply rooted parts control breathing, the sleep-wake cycle, automatic reflexes—the autonomic processes that control the unconscious but crucial life functions. The closer a part of the brain is to the surface, the less crucial its function. For example, the visual cortex is situated directly under the skull, on the outermost part of the brain. A person could live without sight, but not without the ability to breathe.

---

### The Heart-Lung Machine and Heart Transplants

The heart was the first organ to be transplanted successfully as well as replaced artificially. (Back in 1954, Joseph Murray succeeded in transplanting a kidney, but there are still no artificial replacements for the kidneys.)

Heart surgeons sought technological methods by which they could operate on resting, nonbeating hearts. During operations, the heart's functions would have to be replaced by a machine in order for blood to continue to flow through a patient's body and bypass the heart.

In 1937, John Gibbon successfully managed to replace the hearts and lungs of cats with a machine for a short period of time. But, because of the secondary status such research took during World War II, it wasn't until May 6, 1953, that a heart-lung machine was used in an operation on a human being. John Gibbon's machine assumed the functions of the heart and lungs of an eighteen-year-old woman for twenty-six minutes.

At the same time, doctors began inserting artificial parts into living hearts. Today, complete artificial hearts exist. The emotional burden, short duration, and poor quality of life suffered by patients with artificial hearts can still be rather severe.

In 1959, American surgeons Richard Lower and A. Schumway transplanted

(continued)

the first living, functional heart. The transplant was performed on a dog, how-ever, and the dog lived only a few days. Human heart transplants have been suc-cessful since the 1960s.

The first successful operation on a human took place on December 3, 1967. Surgeon Christiaan Barnard transplanted Denise Darwall's heart into the chest of the grocer Louis Washkansky in the Groote-Shuure Hospital of Cape Town, South Africa.

Today, some people even live with two hearts—their own and a second "piggy-back" heart. The first of its kind was given to a little girl named Jacque-line from Hagen by doctors in 1991 in the University Clinic of Münster, Germany. Under Deniz Kececioglu's supervision, the medical team had already performed an operation on a defect in Jacqueline's heart, a defect that had existed since her birth. The defect had weakened the heart to such an extent that Jacqueline could not have survived. That surgical procedure was not enough—the medical team transplanted another heart onto Jacqueline's heart. The additional heart supported Jacqueline's for four years, while her own recovered from surgery, af-ter which the foreign heart was removed.

Nowadays, people between the ages of ten and seventy can receive heart transplants. In 1986, a new heart was even given to a newborn. In general, heart transplant recipients should not suffer from any sort of weakness of the immune system. To prevent rejection of the new organ, the immune system of the recipient must be restrained. Any further susceptibility to environmental in-fluences can result in rejection of the new organ and the person's death. In the first year after the operation, rejection and infection are the two main causes of death among heart recipients. Since about one-seventh of all heart recipients are susceptible to hardening of the arteries during the first year after a transplant, a second heart transplant is highly probable. And the number of second heart transplants is likely to increase over time.

Even when the outermost parts of the brain die, the parts that are more deeply located can function for a very long time. That is why peo-ple who suffer from Alzheimer's disease often spend more than ten years in nursing homes and hospitals. Alzheimer's patients tend to lose their conscious abilities—mostly memory—but the life-critical abilities con-trolled within the deeper brain structures continue to function far longer. Since a person is considered brain dead when all regions of the brain cease to function, all that remains of that person is a body, which exists in the sleeplike state of a kind of suspended animation. The body is artificially supplied with blood, the heart still beats, and oxygen fills the lungs. Detlef Linke, a doctor at the University of Bonn who has even transplanted brain tissue, pictures it this way: "Corpses, thus, can continue to breathe. One could decide whether one wanted to bury them or keep them." And keep them, we might add, for a very long time. Critical care, artificial respirators, and other components of modern medical technology allow the body to be sustained.

In his book *Brain Transplants: The First Immortality on Earth,* Professor Linke sums up the discomfort doctors feel when diagnosing "modern death":

> Up to now, scientists have spoken of "brain death *syndrome,*" though a *syndrome,* a pathological term, is ordinarily ascribed only to a *living* person. Doctors have been forced to view death as a form of pathology! Even physicians have tremendous difficulty accepting the definition of a dead person in terms of brain death.[1]

Not only relatives but doctors may also doubt in their "heart of hearts" that the dead are truly dead. Detlef Linke demonstrates the conflict with an example: "Brain-dead people can show many signs of life that were sometimes even denied them perhaps during life. Such 'deceased persons' often have continual erections. Scholarship on nerve functions will claim that this is merely a reflex of the spinal cord. But what does 'merely' mean?"

## TRAPPED IN A LIFELESS BODY

It is generally agreed that the decisive sign of a human death is the death of the brain. Should we be afraid of one day being fettered to a hospital bed as one of the "undead," with only an artificially functioning body? What if some part of the conscious mind survived extreme trauma and we were otherwise trapped inside a motionless body?

The fear is both justified and not. Justified, because there are cases in which brain and body become separated from one another; the conscious mind exists even though the body may live solely by artificial means. After a stroke that occurs in the brainstem, in the area of fundamental, life-sustaining functions, it is possible for the brain to no longer perceive its body. The medical expression for this is "locked-in syndrome." Only the eyes, fixated on some point in the distance, allow potential access to the brain of the victim. Some people so trapped in their bodies can move their eyes up and down a bit. Today, we try to communicate with these people by considering a movement upward as "Yes" and a movement downward as "No." With the help of an assistant, the French magazine editor Jean-Dominique Bauby, who suffered this traumatic fate, wrote an impressive book, *The Diving Bell and the Butterfly*, about his experience.[2] He spelled out his story with only the blink of one eye.

There are other stroke victims, however, who cannot even move their eyes. These people's brains are imprisoned in the body until death, that is, until brain death. Even though such people show distinctive and normal brain signals, their bodies are dead. Nonetheless, a person who is the slightest bit conscious would never be declared dead prematurely. And so the answer to the question Is existence as one of the "undead" possible? is "No."

---

### Brain Signals and Currents

Brain signals and currents are actually weak electromagnetic waves. Physically, they are like radio or X-ray waves. There is absolutely nothing mysterious about brain signals and currents. The currents' frequency is comparable to that of the currents that flow through electrical wires. The electrical currents originate in the cells of the layer of the brain's cerebrum. We can measure the electrical charges produced by the brain by attaching signal detectors, electrodes, to the skin of the human skull. The electroencephalograph (EEG) records the brain currents and makes them visible on a graph.

When a person sleeps, his or her brain transmits different wave signals at certain predetermined times. The interpretation of these signals is not so simple, since the signals are embedded in many different but simultaneous wave transmissions. With the help of a brain signal conductor and complex mathematics, the precise location of an electrical signal can be pinpointed to within a millimeter in the brain. Nevertheless, individual "snapshots" of brain waves produce rather clear and singular images, each different, for example, for people suffering from schizophrenia, depression, or alcoholism.

Today, it is even possible to use brain signals to control simple machines. As

(continued)

with the EEG, electrodes are placed on the head. The signal they receive is mag-
nified and analyzed by a computer that distinguishes from among three brain
signal patterns, for example: "Person wants to lift the left hand," "Person wants
to lift the right hand," or "Person wants to move the right foot." If the computer is connected to a motor, a handicapped person can control a wheelchair, a
speaking computer, and more. "Studies performed on four test subjects have
shown that a precision of classification of over 60 percent was possible after only
two training sessions, during which the computer learned the particular EEG patterns of the test subject," reports Gerd Pfurtscheller, biomedical technician in
Graz, Austria. Even though they are far from being human, the Graz computers
are fast learners. And work like this has opened the door for the future of brain-
issued control of machines.

Brain death is certain when no more brain signals are measurable and
the person in question no longer has a trace of automatic reflexes. The
pupils no longer shrink from light. The person is no longer conscious,
there is no more evidence of a tracking gaze (when the eyes follow a finger in front of them), the pupils are fixed, and there is a lack of any
bodily reaction to pain. If this state lasts more than twenty-four hours,
the person is considered brain dead. One exception occurs in cases of
poisonings that cause a temporary shutdown, mimicking brain death, of
life functions. Arms, legs, and genitals can still move, since the spinal
cord is responsible for these movements rather than the brain itself. It
can happen that the dead reach out and embrace their caregivers. These
bodies do this without any consciousness or intent. Similar movements
of people who have been hanged to death have also been documented.
The tremors that shoot through the hanged body as well as the release of
urine and excrement are caused not by the brain but by the spinal cord.
The brain is incapable of functioning in these cases, since it receives neither blood nor oxygen.

## HOW DEAD IS A BRAIN?

"No more measurable brain signals," "no unconscious reflexes (as in the
gag reflex),"—yet why is it that, in spite of these seemingly unambigu-

ous indicators, a week doesn't go by without some published article on the subject of brain death? This is because the brain is composed of many parts, all of which have very different functions. Some of the different parts of the brain can die more quickly than others. For the definition of brain death to hold, it is particularly important to distinguish between the cerebral cortex and the medulla. The medulla sits near the base of the skull and is, in evolutionary terms, the oldest part of the brain. Sometimes it is considered the "upper extension of the spinal cord." The medulla regulates the essential bodily functions.

Our consciousness, on the other hand, sits within the cerebral cortex, directly under the inner surface of the skull. The uppermost part of the cerebral cortex is called the neocortex. The neocortex is the result of one of the most recent evolutionary adaptations in human anatomy. In frogs, for example, the neocortex is only about as large as the other parts of their brain. The neocortex is not yet a crucial part of the frog brain. In rats, the center for the sense of smell is rather large when compared with the neocortex. The sense of smell is crucial for these animals. In humans, the large neocortex is what permits higher mental functions, like the ability for language.

Most scientists believe the personal uniqueness of a single individual is based in the neocortex. This is where the definition of brain death can get muddled. "Without consciousness there can be no personality and no person," says Dr. Shann, head of the intensive care unit of the Children's Hospital in Melbourne, Australia. According to him, consciousness alone should determine whether a person lives. "If a person is conscious, he or she is living, regardless of whether that person can breathe, has fixed pupils without pupil reflexes, shows no gag reflex, and no longer reacts to pain," Shann says. "We know that the death of the cerebral cortex is not necessarily followed by the death of the rest of the body because the body with a dead brain can be kept alive for months with infusions of low doses of (the substances) vasopresin and adrenaline."

Nonetheless, some intensive care physicians are greatly concerned with brains that are no longer conscious and that can cause the irretrievable loss of all vestiges of a person's personality. An unconscious brain could cause indirect damage to vital organs, and a new dilemma then presents itself: when can living organs be harvested from brain dead patients? "Organ changes must be absolutely controlled if the organs are to be optimally sustained during a transplant," Doctors Power and van Heerden, Australian intensive care physicians, explain.

Removing an organ when the patient is fully conscious, as was done humorously in Monty Python's movie *The Meaning of Life,* is completely out of the question. Brain death must be confirmed beyond any doubt before the organs are removed. All that is required as proof of brain death is a half an hour of "flatlining" or "electrical inactivity" of an electroencephalograph readout.

When an EEG reading is not possible, say for technical reasons, contrast media or radioactive substances can be injected into the blood flow to the brain in order to determine brain death, the head then being X-rayed. If the brain remains dark on the X-ray, this means no contrast medium or radioactive substance reached the brain. There is no evidence of blood flow to the brain and subsequently, no oxygen supply. Without oxygen, the brain dies within five minutes.

There are other means of determining brain death, for example with ultrasound, magnetic resonance imaging, and computer-assisted tomography. The advantage to these methods is that they produce an image immediately that can be analyzed in real time. While the recording of brain currents or the injection of contrast media uses up valuable minutes and seconds, a doctor has immediate access to the results of MRIs, CAT scans, or ultrasound.

From a purely technical perspective, there is no problem diagnosing brain death. The only trick is to act as quickly as possible. L. H. Monstein, physician at Johns Hopkins University in Baltimore, explains why the new techniques are the most popular: "Image-producing methods allow brain death to be determined earlier, thus allowing the condition of transplant organs to improve and to reduce costs."

There are a few essential rules to caring for the body of a brain dead person so that the potential organ transplants remain as functional as possible. The medical "rule of 100" states, for example, that the body of a brain dead person should maintain a blood pressure of 100 millimeters of mercury, a pulse of fewer than 100 beats per minute, and a urine volume of over 100 milliliters per hour. The "rule of 100" applies to both intensive care patients and brain dead patients. Care according to the "rule of 100" requires giving the patient oxygen and, sometimes, "T3," a thyroid hormone.

While there is a systematic set of medical rules for the diagnosis of brain death, medicine collides with other strongly held beliefs about the nature of death, brain death, and organ transplantation.

As a result, human organs are considered general property rather than essential parts of a person's identity. A terrible conflict! Do we want, or should we be forced, to donate organs to others in order to ensure their lives? Or do our organs belong to our physical and personal identity?

These questions are all the more urgent in light of the following development: people in industrialized nations are living longer and longer. At the same time, surgical methods are improving. As a result, more and more organs are needed for the growing numbers of the aged—especially those once considered beyond hope. There is a significant roadblock to the unbridled harvesting of organs, however. In order to transplant more organs, more blood is required, since transplant operations are a rather bloody ordeal. That there are less and less rather than more blood donors could burst the bubble of the dream of widespread organ harvesting.

Only a small number of physicians are comfortable with the notion that organs of the brain dead are common property. If one agrees with Dr. Shann that there can be no personality or individual identity without a viable neocortex, then the organs of a brain dead person don't belong to anyone. From this perspective, organs are nothing more than material or usable parts that belong to everyone and no one.

---

### Signs of Indisputable Death

To be certain that a person is not buried alive (which never happens in reality), a physician (or, in older times, a well-informed lay person) examined the condition of the body. In addition to the lack of a pulse, the cessation of breathing, and brain death, the obvious symptom of death is rigor mortis. Rigor mortis occurs when the molecule ATP, required by muscles to relax, disintegrates. About two hours after death, the muscles stiffen, starting with the jaw. Once rigor mortis sets in, after approximately six to twelve hours, it is impossible for one person to unlock a dead body from its position. The joints are absolutely unmovable. At room temperature, rigor mortis eases after about two days when the muscle tissue starts to decay.

Reddish spots (called post mortem lividitis) on the corpse are an even earlier sign of death. They appear about an hour after death because blood flowing

*(continued)*

through the smallest vessels under the skin, the skin capillaries, stops moving and "pools." At first, you can push the spots by applying pressure to the skin, but after around two to three days they are immobile—the blood has coagulated.

Although not required for the official diagnosis of death, other obvious signs of decay occur once someone is really, truly dead. Veins appear from just beneath the surface of the skin, decayed and disintegrating from bacteria. They glimmer through the skin as a "net of veins," the ghoulish appearance of which has inspired scores of mask makers—for example in the 1996 feature film *Seven*.

Another interesting phenomenon is the continual settlement of insects on the corpse. Shortly after death, female blowflies lay their eggs in and around the nose, mouth, and ears of the deceased. Whereas maggots usually grow within a few days or weeks, leaving as grown flies, other insects, types of flies and beetles, settle in. Various insects—exactly which ones depend on the stage of decay—feed on the corpse. In fact, specialists can determine from the presence of insects exactly how long a body has been dead—even months after the time of death—which has proven a valuable tool for forensic investigations.[3]

## FOREVER YOUNG: NEOTENY

The Mexican amphibian axolotl stays young forever. Although it reaches an appropriate reproductive stage when it becomes sexually mature, it retains all the characteristics of a young, premature animal. It does not experience any of the usual biological changes associated with reproductive maturing—it experiences a biological anomaly known as neoteny. Such animals look like newborns, others appear to be little more than embryos. Yet they are capable of producing offspring—they become parents without aging. If an axolotl is forced to mature, stimulated by hormone injections in a controlled laboratory setting, it could produce the external effects of a mature individual—like the transformation of a tadpole to a frog. But, in doing so, the axolotl would actually be transformed into another animal entirely; an amphibian resembling an adult tiger salamander would develop. Salamanders and other amphibious creatures also possess the same astounding capability. They start as tadpoles, before they undergo a complete transformation as they grow into adults. In the spirit of the "selfish gene" (whose sole interest is in reproduction),[4] one could say that the sex instinct develops disproportionately and overtakes the body in its development.

Signs of neoteny are clearly visible in humans as well. Several aspects of the human body strongly remind zoologists of characteristics typical among young, immature, even embryonic forms of primates. Among these are the size of the brain, which is very large in comparison to the rest of the body (like an infant's), the angle of head to spine (a right angle), and a mostly hairless body.[5]

Eternally young axolotl amphibians lack the ability to produce a certain growth hormone. Whether one should treat people with a drug working against that hormone in an attempt to maintain youth is rather dubious. Contrary to the axolotl amphibians, who have clearly developed a survival strategy for an appearance of eternal youth, humans who never reach adulthood could not experience fully adult human lives. They would be missing just about everything that would make life worth living. A particular problem would be the sex instinct or reproduction. If the human body is artificially "set" to be eternally "young" through hormone drug therapy (with the goal of living longer), she or he would not develop any reproductive sperm or egg cells. No reproductive cells, no urge to procreate, no sex drive, no further generations: a dead end.

---

### The Shape of Extraterrestrials' Heads

The characteristics of a young child overlap with the common image of aliens: a large, bald head, big eyes, and a comparatively small body. Almost all drawings and falsified photographs of presumed or imagined extraterrestrials have exactly the same characteristics that scientists identify as "neotenous" (young in an otherwise mature body): a large head without hair, not very developed body parts, small hands and feet.

This is, of necessity, the ideal image of an extraterrestrial. If we reject the possibility that these descriptions are based on actual encounters with extraterrestrials, biologists and psychologists must assume that this image is somehow firmly rooted in our psyches.

Psychology lecturer Susan Blackmore from the University of Bristol believes that the image is defined in our heads: "How can healthy, intelligent, and friendly people believe that four-footed extraterrestrials visit our planet and

*(continued)*

abduct humans? All the abduction stories are remarkably similar. They are screaming to be compared with fairy tales and legends, such as wicked witches and hags that creep up on their victims at night and try to suffocate them. The stimulation of certain parts of the brain, together with culturally and individually performed contexts, could be responsible for these similar fantasies. It is not a question of whether extraterrestrials exist, but what these experiences of abduction tell us about our psyches and our brains.[6] It is amazing that the future human being (a new subspecies, perhaps) looks the same as the 'contemporary alien'—a large, bald head, a flat nose, large eyes—a development of humanity in the direction of embryonic characteristics."

The reverse also happens, namely, that some animals grow old too quickly. The images of such people are often the object of shocking reports in the press. In 1994, the French boy Danny was presented by the media in 1994 as "the oldest child in the world" (see figure 10).

Recall the earlier discussion of the very rare illness progeria (Hutchinson-Guilford syndrome), which, when it strikes, results in rapid aging, so much so that hair and teeth fall out during childhood. Rheumatism, heart disease, and glaucoma follow only a few years later. Death comes too early. To this day, the illness is incurable.

Some changes in the genetic makeup that lead to a very similar illness, Werner syndrome (named for the doctor who first described it), were discovered in record time, however. In April of 1996, research groups from the Veterans Health Care System and the Seattle biotechnology company Darwin Molecular reported the discovery among Werner syndrome patients of an altered protein. In contrast to Hutchinson-Guilford syndrome, the illness strikes in adulthood. It leads to premature signs of aging including diabetes, gray hair, and brittle bones.

David Galas, vice president of Darwin Molecular, reported that "the function of the defective gene product is so obvious." Indeed, the Seattle team's discovery affords us great insight into the processes of aging. Werner syndrome is caused by the malfunction of a protein involved in the multiplication of genes. To be more precise, the protein usually unwinds the two strands of DNA so that they serve as the basis for the production of new DNA. When a cell multiplies through division, each daughter cell receives one old and one new strand of DNA. Research

10 Two diseases that cause radically premature aging are progeria and Werner syndrome. Progeria children like Danny look like seventy year olds at the age of nine. The physical decay occurs a bit later in Werner children. The gene for Werner syndrome has been found, thus allowing for experiments with gene therapy. (Photo: France 2)

leader Gerard Schellenberg points out that this protein is responsible not only for the multiplication of genetic material but for its repair as well. The research results from the Seattle-based group lend support and strength to the notion that physical aging could be caused by ever greater damage to genetic materials. This will allow the first treatments for Werner syndrome to be developed.

## GENE THERAPY

Section 1 of this book discussed the DNA-protective caps for genes (telomeres) that wear down in a natural aging process with each cell division. If the wear and tear could be prevented, so to speak, the aging process could be halted. What possibilities are there to conquer aging (and death) with gene technology?

To estimate the possibilities, we must understand the techniques of genetic intervention. The fundamental concepts of this area of biomedical research are rather simple. To change the genetic material of existing cells (especially in the case of a young person or an adult with a hereditary illness), an improved version of that part of the defective DNA would have to be "smuggled in." The problem: a grown body has trillions of cells. And all of them, probably, would need to be so altered. How can this be successful when each individual cell is microscopic? An individual egg cell could be placed under a microscope and injected with new DNA through a very thin glass needle. But one couldn't take each and every sick cell from an entire lung, change them all, and replace every one.

The solution to this dilemma is contained in agents that act as tiny biological "needles," or viruses. The desired genetic information could be fed into a virus, whereupon the virus would be cultivated outside the body and, when ready, injected into the patient's body. There, the viruses would drive new genetic information to the sick cells. A virus is neither alive nor dead and consists of nothing more than a bit of genetic material and a protein shell. Some viruses contain docking ports on their protein shells that can latch onto a cell, allowing the virus to inject its DNA. Not all viruses do this—many appear to melt within the cell's outer layer, releasing copies of their DNA or RNA into the host cell.

In gene therapy, the virus researcher cuts all unnecessary sections from the DNA strand of a virus and replaces these with the "healthy" genetic information of a living person.

It would be ideal if, with a bit of fluid, the virus could be filled into spray bottles. The patient could open his or her mouth, pump the spray, and deeply inhale the viral concoction. That is one possible way in which the viruses can arrive at the place where they take effect, in this case, the lungs. They attack the cells of the lungs and inject the "healthy" DNA into them. This new genetic information is then recognized by the lung cells and converted into proteins. If the lung cells had not converted these proteins before, they can do so now, given the new DNA information. Preliminary stages of this method are already underway. They belong to the area of somatic, or physical, gene therapy. These kinds of therapies are intended to target specific illnesses and are not passed on to children. Since descendants inherit only those genetic changes made in the reproductive system, somatic gene therapy is a remarkably uncontroversial issue. It may make a contribution to the therapy of many incurable diseases that are not otherwise treatable by traditional methods.

Gene therapies are possible in a few parts of the world today and some are even performed in North America. However, they are very costly, labor-intensive, and seldom successful.

In his book, *The Case Ashanti,* Larry Thompson wrote the story of the first somatic gene therapy, inaugurated September 14, 1990, on four-year-old Ashanti DeSilva at the National Institutes of Health in Bethesda. Ashanti suffered from a weakness of the immune system caused by a section of altered DNA (the ADA gene) in her blood cells. This special immune weakness, SCID, severe combined immunodeficiency syndrome, is purely hereditary and has no relation to acquired immonodeficiency syndrome (AIDS). SCID patients must be hospitalized, encased in plastic bubbles, to ensure that all contact with germs is avoided. Before the introduction of gene therapy, a boy named David was the oldest survivor of SCID: he survived twelve years in a plastic bubble at the Baylor College of Medicine in Houston. During a bone marrow transplant that, it was hoped, would save David, he became infected with a herpes virus and died.

Ashanti's case developed differently. A team led by physician and scientist French Anderson applied the following techniques to cure Ashanti: they took the four-year-old's white blood cells, which were missing a very specific hereditary characteristic. The cells were brought into a laboratory in sterile plastic bags and were cultivated there for one week. The researchers built the gene that provides the necessary defense characteristic—which every healthy person has—into viruses they had rendered harmless. The viruses were then combined with Ashanti's white blood cells, whereby the viruses "infected" the blood cells with the healthy gene. Finally, they injected the genetically altered white blood cells back into the girl's bloodstream.

Ashanti's white blood cells had not been able to produce an enzyme— a highly effective substance known as ADA, which is absolutely necessary for resistance to diseases. The treatment was successful to the extent that Ashanti was allowed to play with her friends outside the confines of the hospital and could also go to school. But since white blood cells tend to lose the new genetic information relatively easily, treatment has to be repeated.

In spite of several successes in recent years, gene therapy is still in its infancy. This has to do with the fact that there are thousands of different diseases, each caused by a single altered gene. Gene therapy costs tens of thousands of dollars, thus requiring professionals to determine which hereditary diseases can be treated and which cannot. At this time, gene therapies are performed within the framework of research, so that they do not cost their few patients anything. This will change, however, if gene therapy becomes a more commonplace practice.

The possibilities of a cure through genetic tinkering are limited, unfortunately, not only by high development and treatment costs but also by the low success rate of the treatment. Another point is perhaps the behavior of patients, or rather future patients. In 1996, the German daily newspaper *Süddeutsche Zeitung* printed this headline: "No Interest in Tests for Hereditary Disease." The story: Ellen Wright Clayton from the faculty for medicine at Vanderbilt University had offered a gene test to the residents of Nashville, Tennessee free of cost. The results of this test would allow her to ascertain whether the adults were carriers of a defective gene responsible for a severe metabolic disease, cystic fibrosis. On average, one out of every thousand newborns has this disease. A so-called recessive hereditary illness, it occurs in children born of parents who have both inherited the defective gene. Cystic fibrosis is torturous and fatal because it leads to very congested lungs. Out of one hundred thousand people Clayton addressed with flyers and posters, only two hundred expressed any interest. Two-thirds of those queried expressed anxiety that a positive test result would exclude them from health insurance plans. And Clayton had even worse luck: when she told her small group that a small drop of blood would have to be taken from their finger for the test, another 58 percent refused.

---

### A Short Interview About Genetic Therapy

Professor Walter Doerfler is a molecular geneticist, biochemist, and physician. He has been the director of the Institute for Genetics at the University of Cologne. His work revolves around DNA viruses, medical genetics, and, more recently, the intake of foreign DNA into mammals through food.

*In your opinion, how much further will you be able to go—financially, not ethically, speaking—with the development of gene therapies?*

(continued)

Molecular biology, not just human genetics, already has a very central role in clinical medicine. Diseases caused by several genes simultaneously are not well enough understood to make a prediction regarding treatment. However, monogenetic illness (caused by a single defective gene), although rare, could be possible candidates for gene therapy.

Compared with common illnesses such as heart and circulatory diseases, tumors, and infections, there are really only a minuscule number of hereditary illnesses. On the other hand, these patients are considerably impaired. Something must be done to help them, and something is being done.

Experimental tests into the possibility of gene therapy to treat diseases that arise from only one altered gene are already underway. Perhaps one day they will succeed.

I, on the other hand, am not so optimistic that gene therapy will be broadly applicable in the near future, but it could make an interesting contribution to efforts to cure inherited diseases. I think it is the first step on a long road ahead of us. Gene therapy is also being studied as a treatment for other illnesses such as tumors.

*Your prediction, then, would be that gene therapy has the greatest potential in the treatment of cancer patients?*

I know that many groups are working toward this, but whether it succeeds . . . You know, a lot has been written about gene therapy lately, and certainly many positive things have been said. I would add only this: Dear people, or dear patients, do not expect too much. Gene therapy is only one of several options, yet in its infancy. There will certainly be a number of serious failures as treatment methods are slowly improved. We should really prepare ourselves for decades of development.

The successful forms of therapy that are used today once needed a lot of time before they could be employed effectively. Today, the treatment of tumors is much different from forty years ago, in the beginnings of chemotherapy, when I worked at the medical clinic of the University of Munich. Back then, chemotherapy was quite dangerous. Today, there are treatments whose successes were never anticipated. We are still at the very beginning of gene therapy, the way we were thirty-five years ago with chemotherapy. Maybe in ten or twenty years we will be much further along. One should not promise patients too much too early, because then the entire discipline of genetics comes under, in my opinion, totally unjustified criticism.

*Would your prognosis be that gene therapy will be simply one component of medical technology, or will there be a revolution in physicians' treatment strategies?*

*(continued)*

Both. The revolution has in fact already occurred. Uta Franke, a human geneticist who works in Stanford and with whom I collaborated during one of my sabbaticals, once said, "Molecular biology brought about a renaissance in medicine. Many interesting possibilities have resulted for young researchers through new modes of thinking, new concepts, and new contributions."

I do believe that genetic engineering will incite a small revolution. But that doesn't mean that the world will be perfect tomorrow. Through a slow process, gene therapy will bring qualitative differences to medicine.

*Is it worth spending so much time and money on gene therapy to cure people when there are so many other pressing problems such as overpopulation and the resulting difficulties with energy conservation?*

These problems, too, must be addressed. Energy conservation is a matter for physicists, overpopulation a matter for population geneticists. Overpopulation is certainly a very serious problem, one that obviously cannot be solved, however, by telling sick people, "You have to die."

This doesn't mean that a doctor can necessarily save lives in every case. There are, I think, natural limits in place. But if people who are old and want to live could be given the chance to remain healthy and take care of themselves, that would mean an enormous step forward in medicine.

## DOLLY, THE MOST FAMOUS SHEEP IN THE WORLD

A severed branch can grow new roots if placed in water or damp soil. A branch cut into fifty pieces and placed in small cylinders can grow into several new plants as well. Even a mixture of stems, blossoms, and roots can grow into a plant again under the right lab conditions. Humans who yearn for immortality thus dream of the following possibility: they sacrifice an expendable body part (a toe, an earlobe, some muscle tissue) and try to "breed" it until one or several "copies" are made. These would be stored in a freezer with liquid nitrogen for the day that one of the replacements would be needed (especially if the original was lost or damaged beyond repair).

Why doesn't anyone bring such an idea to fruition? The solution is, biologically speaking, simple enough, but it has serious implications. Only plants and very simple animals are capable of growing an entire body out of their own individual parts. Why? There are several theories, all of which boil down to the same conundrum: a complex body loses the ability to resurrect itself from a puree of tissue; that is the price for its high degree of development and complexity. But several technical problems that had seemed unsolvable in the past were indeed eventually solved using new methods and techniques. We can assume that, ultimately, the reconstruction of a body from random cells will be possible some day. There is, however, already in hand, another technique for producing copies of living creatures. As recently as twenty years ago, only specialists knew the term for this technique. Today it is a common key term, in all languages, associated with gene technology. It is *cloning*.

On February 27, 1997, a research group of the Roslin Institute and the company PPF Therapeutics from the Scottish city of Roslin reported in the journal *Nature* that, for the first time, they had succeeded in completely cloning a large mammal. "Completely" meant that a descendant—the sheep 6LL3, or Dolly—had been born and was living. This time, the researchers had not simply separated very young, fertilized egg cells (after they had divided) into individual cells and allowed completely identical descendants to grow without any further intervention in the DNA. (This method, the separation of fertilized cells that have already divided several times, has been known for years. It has nothing to do with actual cloning.) Instead, the Roslin scientists transferred the genetic information of an adult body cell to an egg cell and used the DNA again from the very first stage of development to "build" the entire sheep clone.

As early as 1977, the nucleus of the skin cell of an adult frog could be made to allow denucleated frog culture cells to grow into tadpoles. Since amphibians and frogs can even regenerate amputated body parts, it was not believed that animals other than tadpoles could be cloned. Furthermore, none of the little proto-frogs managed to live beyond the tadpole stage.

Sheep 6LL3 ushered in a turning point. It survived a gestation period of 148 days and, at its birth, instantly became the most famous lamb in the world (weight at birth: 14.5 pounds). Although researcher Ian Wilmut himself, as well as colleague scientist Colin Stewart were doubtful that Dolly was a true offspring of older, already developed cells, the technique of denucleated cells (the Dolly technique) had been successfully carried out for the first time in mammals. Not all geneticists believe that Dolly is a true clone. Another cloned lamb by the name of Polly, born in July of 1997, could have resulted from an experimental imprecision. In that case, both sheep would be ordinary animals from artificial fertilization. This does not alter the fact that the clonings that have taken place in the past five years are real. In Germany, a law protecting embryos forbids cloning humans: "Whoever creates a human embryo with identical DNA as another embryo, a foetus, or a living or deceased person will be sentenced to up to five years' imprisonment or will be fined." (However, under the impact of recent scientific and economic advances, the law might change.)

Even if the details are highly complex, this type of cloning is relatively simple. The Scottish researchers removed egg cells from a sheep of the Scottish Blackface breed and removed the nucleus from those egg cells. Since the nucleus of every cell contains the complete DNA of the carrier, denucleated egg cells contain almost no genetic information. (In the cell's mitochondria there is a very small amount of additional DNA.) Ian Wilmut's research team united these denucleated egg cells, using the help of a very weak electrical current, with the nuclei of cells from a sheep embryo that was removed from the uterus on the twenty-sixth day of pregnancy or cells from the skin or other tissues of older animals. This technique as such is in widespread use. For several years, a similar method has allowed otherwise immobile sperm cells from men to be united with the egg cells of their female partners (in vitro fertilization, or IVF). In the case of the sheep, the fertilized egg cells were kept in a nutrient solution, or in separate oviducts of sheep, until the blastocyst stage (when the small embryonic mass comprises approximately sixty-four cells) and was then implanted in the uterus of a sheep. Every two weeks, the pregnant sheep was examined with ultrasound. Several other cloned sheep—6LL2, 6LL6, and 6LL9—had to be born by Caesarian section.

All the researchers who participated in this experiment believe that the experiment, technically speaking, could be performed on humans without any great difficulty. The Roslin scientists' work was not undertaken with this in mind but rather with the goal of researching the alterations of the nuclear DNA and the transference of the recorded information it contains while a living creature is developing. All the same, seventeen days before the publication of the Roslin lab results, the molecular geneticist Axel Kahn from Paris asked on *Nature*'s website: "Cloning mammals—cloning humans?" Since Dolly, this question has become highly relevant, not least because more and more frequently infertile couples want to have authentic, biological descendants. "As if that were the only option," grumbled Kahn, who considers old-fashioned adoption the better solution to childlessness. But adoptions could quickly go out of style with the new technique: why choose adoption when a child could be a replica of the father or the mother? Kahn sees his fear realized since observing that "unfortunately, more and more people believe that a person's personality is primarily genetically predetermined." And many people seem to consider the most desirable character traits to be their own. But this egocentric disposition has little to do with the psychological reality: "It is a great advantage to have children who differ from their parents," Kahn writes. "That is why parents love their children as they are and do not try to mold them according to their own ideas. . . . Children are not merely biological but also cultural and emotional descendants."

For that reason, the vision of horror—namely, the technique of producing any random number of copies of one person to achieve political power—that is associated with the future of genetics will probably never come to pass. Power hungry despots could indeed clone an army of identical soldiers that would carry out any and all orders, but the environment in which those clones grow cannot be controlled completely. One could treat these people exactly the same, give them identical meals, show them the same films, give them the same books to read, and so on. It would not work. Cloned people would still experience a range of different emotions and experiences, and each would develop differently as a result.

A leaf falls from a tree at a moment when only one "sibling" is looking. A beam of light hits one person's eye and not another's. Only one nose detects a fleeting scent, causing a memory in only one brain and, subsequently, a unique chain of thought. Fortunately, human beings will always distinguish themselves from others through their varied, serendipitous experiences. Apart from that, it would be far more cost effective to simply steal a number of children and drill them.

Researcher Gunther Stent recently made clear what feelings of envy speculation on genetic possibilities can conjure up and why the value of life is based on its singular and finite nature. He writes, "While it might be delightful to have Marilyn Monroe as a neighbor, it would be an absolute nightmare to meet a thousand copies of her in every corner we look." The very same, I would add, goes for Tom Cruise, Albert Einstein, and the pope.

---

### Clones and Cloning

We call identical copies of a mother organism *clones* (from the Greek for "branch"). If we want to clone a plant, all we need is any body cell taken from that plant. From that cell we can grow a complete new plant that resembles the mother down to the last detail. A mammal, on the other hand, can be cloned only with the help of an egg cell. While the cells of an adult animal can multiply under certain conditions—for example, we can grow a piece of skin in an appropriate mixture of nutrients—an entire living being would never result. This is actually quite remarkable, since every single cell contains the genetic code for the entire organism. But scientists today still do not fully understand why an entire living being cannot develop from any cell of its own body. What is so sensational about the news of the cloned sheep Dolly is that a mature cell was used whose DNA was transferred to a denucleated egg cell.

Biologists have most often used the term *cloning* to mean something else, perhaps better observed as genetic engineering. In this case, cloning implies a technique that alters the genetic material of a living creature.

There are several methods to alter the genetic material of an organism. The principle, however, is the same in each case. A piece of the genetic material of an organism is removed and is replaced by a piece of genetic material from another organism. The new section of DNA gives this organism certain characteristics or abilities it did not have before (see figure 11). The largest area of application for the method is in the production of drugs or immunizations. For example, the genetic code for insulin is inserted into bacteria that then multiply and produce insulin in large quantities. Cloning plays an increasingly important role in farming as well. Useful plants are altered so that they can resist insects, viruses, or fungi, but they are able to survive "pesticide showers" sprayed to rid the fields of weeds. These genetically engineered plants are called GMOS (Genetically Modified Organisms).

---

## ARE THE DINOSAURS COMING BACK?

Cloning humans for the purpose of creating completely identical people fails because our biology does not entirely define our personality. Our personality is not mapped out in our cells but develops as we grow and experience the world. It seems logical to conserve our brains in their entirety, our whole consciousness—instead of mere strands of DNA—for the time when our bodies age, fall ill, and we plan for that second

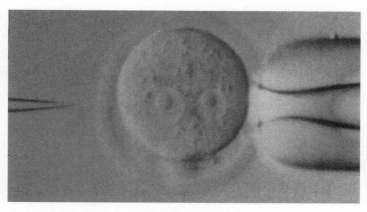

11 Microinjection of DNA into the egg cell of a mouse.

chance, that second life. Nowadays, some people even decide during their lifetimes that either their head or their entire body be frozen and preserved. If one day a cure is found for the disease that might have killed them, their bodies could be retrieved from their icy slumbers and they might be able to continue their lives. In cases in which only the brain could be retrieved, a new body would be needed. The optimal solution would be to "re-create" the very same body, that is, to clone it from its own cells. As just described, we cannot grow a new identical body from a single cell the way we can with plants. The hurdle is the complex structure of a human organism. Such a hurdle, however, might be circumvented, even if it could not be conquered entirely. If some smaller unit of a cell, a single strand of its DNA, for example, could be extracted (and stored early enough), containing the complete instruction set for the entire body, a complete human being could theoretically be reproduced in its original form.

Whether and how this would be possible was the subject of Michael Crichton's novel *Jurassic Park* and Steven Spielberg's film version of that novel. The subject of the sensational best-selling book and movie was not human beings but rather ancient creatures that lived hundreds of millions of years ago. In the imagined Jurassic Park, a modern "theme" park, dinosaurs were reconstructed. Insects fossilized in amber provided scientists with dinosaurs' DNA. The idea behind it: the insects might well have feasted on the blood of living dinosaurs, and the preserved contents of their digestive tracts could contain the DNA necessary to clone actual dinosaurs. The question remains, of course, whether the preserved insects carried dinosaur blood or some other animal's blood. In reality, the experiment would probably fail in its earliest stages. In the book and film, however, the scientists were faced with the problem that the extracted dinosaur DNA was broken in thousands of pieces. Naturally, the fictitious scientists manage to solve this problem. But could this ever really happen?

Up until a few years ago, it was absolutely impossible to entertain even the most remote possibility of the reproduction of dinosaur genes from the existence of old, broken apart DNA. Today, things are different. But first we must answer this question: why does DNA fall apart during the course of time, even when it is preserved in amber, protected from bacterial and other microbial attacks? The answer: DNA, itself an acid, is especially sensitive to acidic environmental influences. Acidic influences are found everywhere, including some soils of the earth, and in the decomposing remains of living things: plants and animals. DNA also breaks apart when exposed to sunlight. Sun, or more precisely, its ultraviolet radiation, does not take an immediate effect, but if a piece of tissue is exposed to light particles for hours, not to mention millennia, the DNA contained within it inevitably falls apart.

Previously, the DNA information strand could be read only when complete, like a book that must contain all it pages in sequence. The genetic blueprints could no longer be read by a single cell from individual fragments of DNA, especially if they were found in severely altered form, as they would surely be in the case of the remnant dinosaur DNA of *Jurassic Park*. Currently, it would take longer than the lifetimes of one hundred scientists to bring the fragments of DNA into sensible order.

Decisive progress in solving the problem of accurate DNA ordering came from U.S. chemist Kary Mullis. According to Mullis, as he was driving through the back roads of California one night he was thinking about a certain kind of molecule stored in large quantities in the laboratory refrigerator at the Cetus company. The utility of those molecules was uncertain, yet they were, in fact, artificially produced fragments of DNA. While on his drive, he hit on the idea that he could use those DNA fragments to multiply DNA using the enzyme called DNA polymerase. In that moment, the polymerase chain reaction (PCR), an incredibly useful technique for biologists and physicians, was born. Today, the PCR technique is in daily use for a broad range of purposes in nearly all biomedical laboratories worldwide. In 1993, Mullis received the Nobel Prize for his discovery of PCR. His method allows vast numbers of DNA fragments to be produced in a test tube.

Why is it useful to mass-produce even more fragmented DNA in the case of dinosaur reconstruction? First, it gives scientists more leeway to try out different methods for reproducing the original DNA. Without duplications, if an experiment failed they would lose their entire source material in one fell swoop. Second, the copies of DNA fragments can be incorporated into other organisms. Their cells then read the DNA and carry out the encoded instructions. This method of cloning was discussed earlier. Ordinarily, the organism used for the incorporation of foreign DNA is the intestinal bacterium *E. coli.*

In the case of the dinosaurs (or human mummies), genetic codes are of little if any use, if they are found scattered among diverse intestinal bacteria. Single-celled *E. coli* bacteria wouldn't be able to handle the complex genetic information of vertebrates. In *Jurassic Park,* the cloning concept was a bit more refined: *several* DNA fragments were incorporated at several places into a *single* strand of the DNA of a living creature. More precisely, a great deal of information about the dinosaur was incorporated into the fertilized egg cell of a living lizard, with the hope that the places on the DNA strand where the dinosaur DNA was patched in at random were well chosen and that the foreign DNA fragments would not be noticed or cause a disturbance to the host organism.

The lizard egg cell would then read all the genetic information contained in the DNA and translate it—the lizard's as well as the dinosaur's. With a lot of luck, a new creature would be formed sharing traits from the lizard and from the dinosaur. Although we know little about the true nature of dinosaurs (e.g., they were probably not the ancestors of modern cold-blooded lizards and are probably related more intrinsically to modern-day birds), the new creature would not resemble anything like the dinosaurs of two hundred or more millions of years ago. The development of the creature would be influenced not only by genetic material but also by the cell tissue surrounding it—by its cytoplasm. All the same, the notion of creating a living monster weighing several tons from a nearly weightless speck of fragmented DNA is compelling indeed. Because of the enormous obstacles in the way, the possibility of realizing such a useless and time-consuming notion remains nothing but a dream. And it is not only the lack of money and time that hinders the resurrection of the dinosaur.

Developmental biologists know that embryos carrying the genetic information of two different organisms would show fatal physical changes and die an early death. Such creatures would be capable of reproduction, in theory, or could become so as adult animals, but, in reality, they would never get that far. They wouldn't have a shadow of a chance to live much longer than a few days. The complex collusion of all cell components during an animal's growth could hardly make use of the information from compiled DNA. With a *lot* of luck and good instincts, we might be able to bring together DNA fragments from very similar animals. For example, the now extinct quagga was a species of zebra that had no stripes on its hindquarters. If we were to take the information that dictated "no stripes on the hindquarters" from preserved quagga tissue and place it in a living zebra's egg cell, a normal zebra could conceivably result with stripe-free buttocks. But enormous dinosaurs and living lizards are so dissimilar that even a preliminary experiment using only one single fragment of dinosaur DNA would in all likelihood fail.

The enormity of the problem of DNA fragment transfer was demonstrated when, in 1986, scientists from the California company Genentech attempted to produce a genetically altered pig. They injected a small amount of foreign genetic material from a cow into eight thousand pig egg cells using a hollow needle (cf. figure 11). The fragment was then supposed to be present in each cell of the adult pigs that grew from those egg cells. The rate of success was devastating: from the eight thousand injected fertilized eggs, no more than forty-three developed into pigs with (slightly) altered genes.

## TREES OF LIFE

One of the people who knows the most about the possibilities and difficulties of old DNA is Svante Pääbo of the German Max Planck Institute for Evolutionary Anthropology. He works with very old genetic material that he gathers from mammoths, mummies, and other preserved tissues. Pääbo became known to the general public in 1985, when he successfully duplicated DNA fragments from the skin of a child mummy of the twenty-seventh dynasty from the Egyptian Museum in Berlin. Pääbo's goal is to study parts of the old genetic material, building block by building block. Scientists call this "sequencing." Its purpose: the more strongly the order of the building blocks of one animal resembles that of another animal, the more closely the two are related. Many of the uncertainties in our current understanding of the evolution of life could be clarified by way of this method.

For example, scientists would like to know how similar dinosaurs and the remains of extinct birds are at the DNA level. Beyond that, zoologists and biologists like Pääbo are interested in the relationship between extinct and living animals. Are mammoths and elephants really as similar as they appear on the outside? And how different were mammoths in different parts of the world? Mammoths were spread out across Africa, Eurasia, and America. Pääbo's material for the experiment won't run out: he believes there are hundreds, or even thousands, of mammoth remains all over the world.

If we compare parts of the (mitochondrial) DNA of about 10 to 20 of these mammoths, we would be able to make our first conjectures about the genetic differences among them. Pääbo's research team has already been able to study four mammoths successfully and with reproducible results. Considering the poor condition of the old DNA, this is a tremendous accomplishment, demonstrating technical dexterity and a great deal of experience. The results were not as plentiful for the mummies from the Nile valley: only 2 out of 110 mummies produced utilizable DNA-sequences (their usefulness is determined by whether several experiments reproduce results). Pääbo believes that the essentially well-preserved mummies' DNA was damaged by the warm climate. Freezing temperatures and concealment from air are clearly the best conditions for preserving old DNA.

Modern DNA, for example from the blood cells of living humans, can be sequenced comparatively easily: the DNA strand from live cells does not fall apart as easily. On the other hand, stored DNA ages. After a while, only pieces remain of the thin acid strand. If forensic scientists find it almost impossible to study the DNA of a ten-year-old corpse, imagine how difficult it must be for Pääbo's forty- to fifty-thousand-year-old mammoth tissue from the Siberian permafrost or his three-thousand-year-old mummies.

The DNA pieces obtained from old mummy tissue are up to three hundred DNA-building blocks long. This is often sufficient to calculate phylogenetic trees of organisms, but it is almost impossible to reconstruct the entire DNA of an animal or human being in its original order. This will remain science fiction for a very long time.

Svante Pääbo confirmed this fact. When I asked him to "answer simply whether it would ever be possible to hatch little dinosaurs in some form or to 'reawaken' extinct plants?" Pääbo's unambiguous answer was, "No. Is that simple enough?" *Jurassic Park* will be what it always was: a modern fairy tale.

## SALIVA DONATION FOR ETERNITY

Perhaps one day, we humans will be able to reproduce ourselves from our own ribs (or nose, if things happen according to a 1973 Woody Allen film). But even people today should not be discouraged. Essentially, it is enough to leave a small amount of DNA behind. This could then be used to reproduce its original owner, if the opportunity arose. The DNA could be obtained from a dried drop of blood or saliva and be preserved in its entirety—in a small plastic container, no larger than a fingertip, in a freezer, or on a specially prepared piece of filter paper— until the desired hour. (Kary Mullis, of PCR fame, has been trying to sell clocks through the company Stargene, the faces of which contain capsules of DNA from famous personalities like Albert Einstein and Marilyn Monroe.) The cost of this kind of preservation is far lower than the cost of preserving an entire body in liquid nitrogen.

---

### Freezing Corpses

By the beginning of 1995, as many as twenty-eight people in Phoenix, Arizona had their bodies preserved in freezer compartments. To be precise, eleven people were completely frozen at this time, while seventeen others only put their heads on ice. The company, Alcor, that offered this service charged $120,000 for the former, $50,000 for the latter, and foreign customers pay an additional fee of $10,000. Most of the customers who are not yet in Alcor storage have bought life insurance that pays for this expensive, afterdeath expensive freezing scheme.

The mobile task force that waits at the bed of the dying for the critical moment was lead by Tanya Jones. She moonlighted as a science fiction writer and worked for a while for the U.S. Department of Defense.    (*continued*)

A few minutes after clinical (complete) brain death, the deceased is placed in a plastic tub of ice, with running ice-cold water. In the meantime, the chest cavity is lifted and lowered by a machine to continue circulation. Ms. Jones then cuts open the trachea and inserts a tube to pump fresh oxygen into the lungs. At the same time, the corpse is attached to a drip with a sugary nutrient solution that flows into its veins. Finally, a femoral aorta in the leg is opened. A blood pump exchanges the blood with Viaspan, a blood-substitute solution ordinarily used to keep donated organs fresh for up to one day. The entire process takes about five hours. Once the body is filled with Viaspan, it is cooled to a temperature of about 35° F. In a box filled with dry ice, the body is then brought to Alcor's storage space.

The fact that these bodies are, in clinical terms, considered dead comes as a source of relief to Alcor's chief, Stephen Bridge. None of Alcor's employees are physicians and are thus not permitted to handle living patients. However, they can saw apart, pump, or freeze cadavers without fear of reprise. "Of course we are very grateful for the possibilities offered to us by physicians," says Bridge. And what's more: if the patients weren't dead, their life insurance would not pay their fees (cf. figure 12).

Stored at Alcor, a customer's thorax is cut open, the lung's blood vessels are separated from the heart, and the brain is connected directly to the heart. The Viaspan liquid in the patient's veins is then replaced with the antifreeze glycerin—six more hours go by. During this process, the brain detaches itself from the skull, since it becomes dried out. If only the head was paid to be frozen, it is detached from the rest of the body, the remains of which are cremated. If the entire body is preserved, it is sewn up, laid in a bath of ice and oil, where it soaks for a day and a half. Finally, it is slowly immersed in liquid nitrogen, where it remains indefinitely at a temperature down to -328° F.

The great irony of this procedure is that, in the long run, DNA can become torn to bits at such extremely low temperatures. In the end, the reconstruction of viable DNA can become impossible—and we are right back to the dinosaur DNA reconstruction problem. It doesn't matter if one person chooses to be frozen (after death) for $120,000, or if another prefers to have his DNA extracted from his blood and refrigerated in a tiny container for $100, or if yet another person has her complete DNA sequence saved on a computer diskette. No one will ever be reconstructed in the future using any of these methods.

And these are just the biological facts. Hans Mohr, Stuttgart professor of biology and board member of the German Academy for the Evaluation of the Consequences of Technology, addressed the ethical and moral questions of freezing corpses in the magazine section of the *Frankfurter Allgemeine Zeitung*: "That is

(continued)

a perversity which I cannot abide and possible only in a culture where excessive wealth and horrifying poverty clash on a daily basis. First of all, such a biological feat can only be accomplished in Hollywood movies, and second, in a country where every tenth child is malnourished, such experiments cannot be morally justified. . . . Such blasé fiddling with death speaks volumes about the spiritual health of a society."

A complete strand of DNA can be transferred to an egg cell at any time. In contrast to the dinosaur DNA, this strand does not need any additional information from a similar organism—it contains all its own necessary blueprints. Instead, the DNA is removed from the nucleus of one cell and is replaced with the complete strand of preserved DNA from which a human could be reconstructed. It makes the most sense to use a human denucleated egg cell, but, in a pinch, a chimpanzee's egg cell might suffice. More precise information than this really isn't known, but the experiment has been completed successfully on several occasions with smaller animals. The only disadvantage to the costly freezing method (of saliva or blood) is that the person's "new" person would have to be "raised" again from scratch. As described earlier, however, the same person would never result twice—its spirit, or psyche, would be different.

How could we preserve someone's personality for a second life? To answer this question, we will address another sensitive issue in the next section, an issue that has certainly occupied the fancies of those who simply cannot come to terms with death.

### BRAIN TRANSPLANTS: UTOPIA OR REALITY?

The death that causes us the greatest anxiety is the death of our brain, our consciousness, ourselves. The cerebrum, the main locus of one's personality, suffers irreparable damage from oxygen deprivation within just a few minutes. Sight, speech, thought processes, and especially memory are permanently lost if this part of the brain is irreparably damaged. At that moment, we stop being ourselves; our soul is gone. Deeper regions of the brain can keep our body alive. Couldn't a new brain be incorpo-

12 Deep freezing of corpses. By the beginning of 1995, the corpses of twenty-eight people were already stored in deep freezers in Phoenix, Arizona. Alcor, the company offering this service, charges $50,000 to $120,000; foreign customers pay an addtional fee of $10,000. (Photo: Michael Montfort)

rated into such a body? What would happen if an active brain were removed from a dying body? The notion of brain transplant is unthinkable for many people. Yet, in principle, there are no real technical problems that would prevent such a procedure. Although the cost would be high, such a procedure could happen within the foreseeable future. In the United States today, brain tissue from aborted human embryos is sometimes transplanted to the brains of people suffering from Parkinson's disease.

When physicians reflect on the possibility of a total brain transplant, two questions are most important. First, is it correct to assume that brain death means the death of the whole person? How can such a clear line be drawn between life and death, even though parts of the brain are still active after brain death and the body can still move occasionally? And, second, who would be a brain donor? Let's assume a sick person and the victim of a sudden accident both possess some living, viable brain tissue: should the Alzheimer's patient (to use an example) give the victim of a motorcycle accident (to use another example) those parts of his or her brain that are still healthy—or vice versa? Which of the two is more deserving to be kept alive?

Whose soul is preserved by an exchange of brain matter, and whose soul is lost? Consider the following thought experiment made popular in philosophy by Alan Shewmon: a person agrees to a medical experiment. In the operating room, the person's brain is removed. Finally, there is a brain at one end of the room, at the other end a person without a brain. Both parts, brain and body, are medically alive, attached to machines, that keep blood circulation and oxygen supply going. Where in the room is the person?

The usual answer is: the person is his or her brain. Today, it is, in fact, incontrovertible that all abilities considered "human" issue from the brain. Recently, this has been most clearly demonstrated by the New York physician Dr. Israel Rosenfeld, who insists that consciousness is an absolute condition of being human. More precisely, consciousness comes into being through the interaction of nerve discharges within the brain. The body itself is not a necessary condition for a "conscious person." Except in the case of life-threatening illnesses, those that might influence brain activity through some deprivation of oxygen, humans experience their bodies as a phenomenon separate from consciousness.

Even on a theoretical level, a living person without a brain is impossible. From this, we tend to deduce that other vital signs, even blood circulation, are less crucial than brain activity. From this perspective, the body is a mere appendix to the brain. In earlier times, the heart rather than the brain was considered to be the most essential part of the body. In fact, for quite some time, the presence of a heartbeat was used as a litmus test to establish whether someone was alive or dead. The brain took over the heart's function in this respect only in the last few decades. In 1950, a California court ruled (in Thomas vs. Anderson) that death begins "when life ends, and that happens when the heart stops beating and breathing ceases."

Today we believe we know better. A person's own brain could conceivably be saved and "made younger" by a transplant into a newer, healthier body. If this procedure were repeated frequently, a new way forward toward immortality would be established. In principle, though, it would not be critical to transplant the entire brain, for, from a medical perspective, only one part of the brain—the cerebrum—is considered to be the locus of the soul.

All stimuli, for example, the sight of a horizontal line, effect a weak electrical current in a limited, more or less predictable part of the cerebrum. The same goes for hearing, the interpretation of what is heard, speech, the formation of thoughts, and so on. Knowledge of the place of the electrical current does not say anything in particular about the feeling associated with it, but there is a clear relationship between the stimulus and the location in the brain of that stimulus. This has been studied and confirmed in countless experiments by brain researchers. In sum, the brain creates a foundation on which these stimuli function and are interpreted, or, rather, they provide the setting for these functions. Who or what else could be the recipient of the jumble of environmental stimuli transformed into electrical activity across the landscape of brain matter?

The only musings about an additional final recipient come from the famous neurologist John Eccles, discoverer of the transference of signals among nerves. Since his ideas are still quite vivid and widespread today (although rather outmoded), I will describe them here in brief. They demonstrate, on one hand, that the locus of consciousness must be in the brain, and, on the other hand, they show that the idea of a functioning "soul," separate from the brain, is an absurdity and an impossibility.

Eccles used the existing, tiny electrical currents of the cerebral nerves as a departure point. The discharge of these currents takes place extremely rapidly during every moment of a person's life, are dispersed well beyond the cerebral cortex, and cannot easily be "caught" in technical terms. For example, the sight of a chessboard does not stimulate a chessboardlike pattern in some unknown part of the brain. Then how do we recognize the chessboard as such? Eccles, who won the Nobel Prize for medicine in 1963, always helped himself and his listeners with the image of tiny lamps that light up in a certain part of the brain with every electrical discharge. Each tiny lamp corresponds to a subunit of the brain, each of them composed of approximately four thousand nerve cells. With this image, we arrive at a sum of about 2 to 3 million lamps per brain. (The subunits of the brain, each containing four thousand cells, were discovered by radioactive markings of the outermost brain layer.)

Observed from a considerable distance, such a brain composed of tiny lamps would give off an incredible changing pattern of light. A momentary glance at this sea of light might remind us of the Manhattan skyline at night. Sir Charles Sherrington, the famous English "philosopher of the nervous system," described the images like this: "The brain is a mysterious loom on which millions of brilliant shuttles weave a confusing but always meaningful pattern that is transitory, of course."

According to Eccles, a "self-conscious spirit" analyzes the pattern of lights. This spirit can recognize, analyze, and interpret the rapid changes in the pattern, e.g., the information "chessboard." The self-conscious spirit must not only recognize patterns but must give them temporal cohesion as well. This ability is especially important, since time possibly occurs only in our heads. Is this spirit then a kind of authority separate from all things? Hardly. The spirit is simply the totality of all nerve discharge patterns found in the brain and, as such, should theoretically be transplantable.

Yet what is technically possible is far from ethically justified. This discussion can only outline those possibilities within medical technology, which would be at the center of such a debate. In U.S. storage spaces, there are indeed frozen brains of deceased people, waiting for the day of their resurrection. If no ice crystals form inside the brain cells, perhaps they will be able to be transplanted into brainless bodies one day. Would this really be without medical problems?

## OLD BRAIN, NEW BODY

Today we can merely speculate on the difficulties of a brain orienting itself in a new body, since our knowledge is still very limited. The most obvious problem is probably also the most uncomfortable. The new body has a different anatomy than the one the brain is used to. No body is the same as any other. (Even identical twins differ, since different experiences affect the body differently.) Two bodies also have, in all probability, differing sensitivities to sensual stimuli. It is not so much the distribution of the stimuli-receiving nerve endings on the surface of the body as much as the structure of the brain that receives and analyzes signals. A brain "trained" in a certain body would probably be limited in its capability to readapt, that is, to meaningfully process the sense impressions in other parts of the brain. Precisely this dilemma could plague a transplanted brain. It is highly questionable, for example, whether a brain could control the movements of a foreign body straight away, since the information given by two bodies could be considerably different—each tiny affliction, every typical motion, every facial expression repeated over the course of years is not only dependent on the architecture of the brain, the detailed structure of the brain is also shaped by how often a certain motion took place and which nerve connections were established. If sense impressions suddenly reach the brain with different strength or form, severe problems can result. If the sense of touch is too refined, it can become a plague, if it is too poor, one would bump and cut oneself without noticing. If taste impressions are transferred to sight regions of the brain, we would arrive at mixed-up sense experiences. E.g., form or sight perceptions can then be coupled with the perception of numbers (i.e., every number is directly coupled to a color and perceived as a color).

In moderation, such changes in perception could be pleasing, or at least interesting, but if they become a continual source of disturbance, those affected suffer greatly. The complete disappearance of individual perceptions is an enormous, depressing loss, as Edwin Morris, a specialist in aroma therapy, describes, citing a specific case in the *New Yorker*. The patient in question had suddenly lost his sense of taste: "Of course I had to eat—I didn't want to die. But most of what I ate tasted like cardboard." Would we have to assume, then, that a brain in a foreign body would not be able to orient itself and that sense perception would be disturbed? We do not know and are not in a position to find out at this point. But there are experiments done on animals in the early eighties that seem to suggest that, at least partially, a brain might at least partially be able to adapt itself to a new body.

At the time, brain researcher Michael Merzenich, who now works at the University of California in San Francisco, studied monkeys that had an individual finger amputated. As he discovered, those parts of the monkeys' brains responsible for the amputated finger do not merely fade away but take over new tasks.

At the same time, the nerve researcher Tim Pons came to the conclusion that the brain is a highly adaptable organ. Ten years after a group of animal rights activists kidnapped his colleague Ed Taub's laboratory monkeys with artificially destroyed nerves, Pons could study the animals once more. As it turned out, those parts of the brain that had received stimuli from the hands and arms, now—after the destruction of the once functioning nerves in the spinal cord—recognized when their faces were caressed. Monkey brains adapt themselves continuously to new bodily conditions and redistribute, for example, the controls for the limbs, if that is necessary. We don't yet know if this would also be the case with people, especially not once they are aged.

## SENSE PERCEPTION TAKEN LITERALLY

Our sense organs are the result of a great evolutionary ability to adapt to our environment. If we say, when recounting an uncomfortable event, that our "hair stood on end," we unwittingly refer back to a time when our ancestors literally made the perimeter of their bodies larger to intimidate the enemy with feigned muscle mass. Eyes and ears, nose and mouth, receiving stations for touch, hearing, balance, smell, and taste, are ideally adapted to the nature of the physical stimuli that reach us. Light alters specific molecules on the retina of the eye that are later transformed back into their original condition. The ear has a snail-like channel equipped with tiny hairs for hearing. Sound waves bend the fine hairs to the side in areas where the pitches are determined through varying degrees of the softness/pliability of the ear tissue or, rather, through the breadth of the ear trumpet. In the eyes, a chemical change occurs while the ears first experience a mechanical change in the form of a bend. Both types of stimulus reception are the basic ones that function throughout the body and go into action equally when a person touches, tastes, or smells. There are great differences, however—e.g., between the eyes and the ears—in their connection to the brain. While information from the eyes can merely be sent in the brain's direction, the ear can receive nerve signals from the brain. Here again is another example of the complicated network of nerve connections and effects that could pose tremendous difficulties for a new, transplanted brain. The receiving stations (receptors) for the senses are dispersed throughout the body. There are sensors located on the surface that perceive pressure, coldness, or warmth, others that feel the slightest movement of individual hairs on the skin, and still others in the mouth that take in different kinds of tastes, whether sweet, sour, or spicy. Inside the body, other receiving stations measure the stretching of muscles, the so-called internal aches (a not specifically localizable pain deep beneath the surface of the body), the position of the joints (to be able to give the body the right posture), among many other things. These perceptions may differ greatly from body to body: the so-called sensory nerve connections in the body of a distance runner, an arthritic smoker, and a brooding poet have adjusted themselves very differently during the course of each of their lives.

In spite of the flexibility of a brain in its original body, we cannot stop or reconcile the basic phenomenon that the brain's nerve cells stop dividing after a person turns fifteen years old. The weight of the brain decreases continuously, and anywhere between fifty and one hundred thousand brain nerve cells are lost daily. With an original number of about 14 billion, however, this brain cell loss during a normal human life is insignificant. During the course of a lifetime, the three hundred thousand miles of nerves in the cerebrum provide more than enough room for information. But the case of a continually transplanted brain could prove to be quite different. When nerve cells disappear, the value of the brain and, correspondingly, the potential quality of life diminishes. Programmed cell aging and death seem to thwart our calculations for a successful brain transplant. Thus it is with good reason that, until now, only small fragments of fresh brain tissue from aborted fetuses with developmental potential are transplanted to partly altered brains of adults who suffer from Parkinson's. This treatment was used in the United States, for the most part, where it was carried out in accordance with very strict rules and regulations.


## LIVE CELLS AS FOUNTAIN OF YOUTH?

The legal situation in countries like Russia looks quite different. In April of 1994, a German pediatrician and a U.S. plastic surgeon created a lot of talk. In a side wing of a Russian clinic they called the "International Institute for Biological Medicine," they offered abortions free of charge. Thousands of women without any means used this opportunity in a desperate situation. The dead embryos (which were up to seven months old) were frozen, processed, and then injected into patients all over the world as live cells for five-digit fees. According to the physicians, the treatment was even supposed to help prevent diseases such as Down syndrome. In truth, Down syndrome, like most genetically determined illnesses, is incurable. This case demonstrates that people's faith in miracles creates a sufficient demand for false cures that is exploited by profiteers.

The fresh cell method was originally developed by the Swiss physician Paul Niehans. He treated two thousand aging men and women, most of whom subjectively judged it a success. Since some famous people were among those treated, the procedure quickly attracted a lot of attention. Konrad Adenauer, Winston Churchill, Pope Pius XII, and even Fidel Castro were customers of the Swiss doctor. Niehans was concerned first and foremost with fighting mental weaknesses caused by aging. Never did he claim to slow down the aging process with live cells.

The following concept lies behind the live cell treatment method: embryonic cells contain a large number of growth factors. These substances, which were mentioned in the first chapter, activate bodily cells to multiply and maintain themselves. The body needs them to develop from an immature organism into an adult.

Theoretically, the growth factors can rectify a primary problem of the aging body programmed to let all its cells die gradually or—more important—not to replace used cells. The skin wrinkles because the cells beneath it go slack; hair turns gray because hair pigments are no longer produced. If the costly cells injected into the muscles of willing patients actually stimulated body cells in an aging organism to grow, this would mean the certain death of the patient, since the random cells stimulated to grow would include undesirable ones as well. The fact that no patients are hurt by the procedure is testament to the fact that, in regard to the desired goal, these shots are entirely ineffective. Medically speaking, there are legitimate objects to live cell therapy that have nothing to do with growth factors. In 1987, the federal public health department in Germany banned the use of dry cell preparations and simultaneously announced severe reservations about the use of live cells, which is, moreover, unrecognized by orthodox medical practitioners as a treatment method. Most recently, several cases have been reported in which the giving of foreign cells or cell parts led to infectious and, in some cases, fatal diseases of the nervous system. On the other hand, allergies to live cells appear astoundingly seldom.

Daniel Rudman, an internist from Madison, Wisconsin, has developed a more polished method of live cell therapy. His team injected people over the age of sixty with purified growth factors from cell cultures. After about six months, the treatment actually showed a positive effect: the skin became thicker and smoother again. Other areas of the body can supposedly be kept young with other kinds of growth factors as well.

The live cell or growth factor treatment cannot postpone the genetically predetermined point of death. It simply helps the body remain vital and fresh within its genetically defined lifespan—a pleasure affordable only by the rich.

A brain transplant—if there should ever be such a thing—remains pie in the sky, and normal live cells are ineffective. This is not at all the case with organs such as kidneys, hearts, livers, and the pancreas. With these organs, the medical and technological obstacles have for the most part been overcome, and each year donated organs lengthen the lives of many thousands of people. Nevertheless, the area of organ transplant is anything but unproblematic. Let's take a look at several difficulties that arise when people want to extend their lives with the help of foreign organs or—in more general terms—with the help of foreign tissue.

## WHO OWNS THE PRESIDENT'S LIVER?

Like many other people, even the president of the United States might have wondered if, after his death, he would like to help another lengthen his or her life with his organs. No one will try to meddle with such an important decision. The freedom to chose is guaranteed to absolutely every citizen of most democratic, Western countries.

But is this really the case? Several critics of modern biomedicine worry that doctors are less and less concerned with treating an individual and focus their energies more on "life in itself." The president's liver could thus one day be considered part of the public domain as soon as he no longer needs it—someone else might need it urgently. In this spirit, a letter was written by Hamburg internist Doris Saynisch to the *Ärzte Zeitung* suggesting to readers that artificial respiration is nothing but a postponement of death in the interests of patients waiting for (and dependent on) a transplant. In other words, a person's organs belong to the organ receiver as soon as the former is brain dead.

Due to the high demand for donor organs, a law was considered in 1996 in Germany that would have made every person a potential organ donor unless explicitly forbidden by the individual. Although this special regulation was rejected in the end, it might only be a matter of time before it achieves legal legitimization: not only the poor ratio of availability to demand speaks in its favor (out of about nine hundred thousand deceased people in Germany each year, only about five thousand—according to the most conservative estimates—become donors for the heart, liver, lungs, pancreas and intestines). Most of all, the impressive, moving reports of people who have been saved by organ transplants and can now see, walk, or live on among their loved ones would have to guarantee a legal regulation morally.

As long as an individual can forbid having his or her organs removed upon brain death, the notion of humans as living organ carriers or keepers will remain on the border of what is acceptable. Formulations like those of the Australian philosopher Peter Singer, who maintains that organs are the "raw materials of health" and should be distributed accordingly from sick people to younger ones, are highly questionable. In the end, it is entirely unreasonable to expect that sick people should be medically "used" against their will, but precisely this could possibly come to pass in the near future, depending on how the so-called Convention on Bioethics is rendered in various countries.

The German left-wing publisher Erika Feyerabend is concerned that, with the European bioethical recommendation, the idea of the "health of the people" would be given priority over the goal of the well-being of every individual. Research on humans then serves the primary purpose of the health of the general population—the individual has to put his own demands on a back burner. In extreme cases, this could lead to a situation in which critically ill people could be examined and have tissue, blood, or bone marrow extracted without their consent. An absolute premium placed on the health of the general population could justify such intervention in corporeal self-determination with the claim that such people have a responsibility to society. Through their donations, they would also help us understand their illness and help us find treatment methods. A similar sort of moral pressure could also be exerted on healthy organ donors. The German Association of Transplant Centers has therefore created its own code of practice: only when a dying person has an organ donor ID or when the relatives give their permission can organs be removed. Only in exceptional cases, between closely related people, are donations from the living permitted.

---

### The Convention on Bioethics

It was actually a good idea: with a Europe-wide agreement on ethics, it was hoped that the misuse of biomedical research would be practically prevented by law, in that certain areas of research would be clearly permitted and others explicitly forbidden. Of the thirty-nine member states presented with this agreement by the European Union, only Spain, France, Italy, Portugal, Sweden, Finland, and Romania could bring themselves after much hesitation to sign it on April 4, 1997. United Kingdom, Germany, Russia, and Belgium voiced their rejection from the outset. Since only five signatures were necessary, the regulations of one of the most controversial documents in Europe went into effect. Germany's own stricter laws remain in place.

The Convention on Bioethics maintained that research on human genes is only allowable when such studies serve the purpose of identifying and preventing illness and when the genes of possible descendants would not be affected. The determination of gender in the unborn is forbidden, except in cases when it could help recognize a severe hereditary illness. Many Germans consider the most important rule to be the one that allows the biomedical examination of people who cannot consent to medical intervention (i.e., children, the senile aged, and the severely mentally handicapped). While this regulation of the rights of those in particular need of protection was valued as a progressive step throughout Europe, German politicians believed that it unwittingly opened a back door that would indeed allow genetic testing in special cases on such people. For that reason, Germany rejected the Convention on Bioethics.

Today it seems that the Germans did themselves a disservice through their position: the most unpopular tests for genetic illnesses that health insurance companies like to require, such as breast cancer, cystic fibrosis, Huntington's chorea (St. Vitus's dance), and Creutzfeldt-Jakob disease, could be pushed through in Germany, whereas these might have been prevented by the Convention on Bioethics. The reason: the Convention on Bioethics would have regulated these tests, while in Germany they are now quite unregulated and are thus freely managed. It is still assumed, though, that Germany's strict laws restrict such testing even better than the convention could in its current form.

*(continued)*

The story of the Convention on Bioethics goes back to 1991, when the European parliament demanded an agreement that would ensure, in light of the "rapid progress in medicine and biology," the "protection of human rights in the realm of biomedicine." While the first draft of the convention was rejected by parliament on the grounds of imprecision and legal loopholes, the second draft in the fall of 1996 met with greater approval. "It was abundantly clear," reported Swiss commission member Gian-Reto Plattner, "that the German delegation, strongly influenced by its history in this century, would continue to have grave reservations in this debate about medical intervention and transplant research on nonconsenting people. . . . Nevertheless, the convention had to strike a difficult compromise with regard to research on embryos that would enable an agreement among members of parliament throughout Europe in spite of their many cultural and religious differences." Indeed, according to the technical rules of the convention, human embryos may be used for research within a fourteen-day limit, if they are "left over" from preparations for artificial insemination, that is, not implanted in the uterus. In connection with the potential of examining nonconsenting humans and possibly withdrawing blood for gene research, this allowance set off a wave of protest from many organizations, including those representing the disabled as well as International Doctors Against Atomic Warfare.

In the October 1996 issue of the *Ärzte Zeitung*, Helmut Laschet demonstrated that, in addition to the entirely justified, sober objections to the agreement, there was a good bit of self-righteousness and hypocrisy as well. Why, Laschet asks, doesn't the unconditional protection of embryos demanded by the Germans apply to embryos that are aborted daily? Why is it forbidden to change the sex of a child to prevent a hereditary illness, while a child who actually has a hereditary illness can be aborted? Why is it not permissible to use the countless embryos for research that accumulate during preparation for artificial insemination, while artificial insemination itself, which brings about these embryos in the first place, is allowed? Why can medications be tested on children and not confused old people? Biomedicine turns the whole world of medical technology on its head and demands ethical and moral debate. A stubborn "no" cannot answer many of the questions it stirs up.

## NONBIOLOGIAL PROBLEMS OF ORGAN TRANSPLANTS

One of the most well-known bioethicists is the Australian philosophy professor Peter Singer from the Center for Human Values at Princeton University. He believes that not every human being has an unconditional right to live. Peter Singer is thus a highly controversial figure, so controversial, in fact, that he has had to cancel numerous planned lectures and has frequently had his lecture invitations cancelled because of public pressure.

Singer is in agreement with other philosophers that ethical regulations should exist, since no one enjoys suffering. He also agrees that the duration and degree of suffering should be as short as possible and pleasure as long as possible. This view is the essence of the theory of utilitarianism, a theory considered petty and bourgeois by Karl Marx. "Petty," in this sense implies a necessity to calculate the relationship of cost to usefulness, to determine the ultimate value of an action.

In Singer's eyes, it is worth keeping people alive. People, however, who "only" live but, because of severe brain damage, have no memory, no ability to make judgments, or, in some cases, have no consciousness, are not "people," according to Singer, but "mere" living creatures. Thus he draws a distinction between "humans' and "people": people seek pleasure and avoid pain. Mere living creatures don't have such interests, for they have no interests at all—they don't even perceive themselves as beings. If they had a single interest, says Singer, it would be the interest of mitigating or minimizing suffering. Since severely handicapped people ostensibly spend the overwhelming majority of time in pain and suffering, they can be euthanized in principle if conditions made this necessary. Euthanizing them would end both their suffering and the suffering of their relatives.

Insanity? Nonsense? Inconsequential blather of one individual? Absolutely not. Some people think this kind of weighing out pleasure and pain against each other defines the modern biological sciences. The social critic Ivan Illich even insists that contemporary medicine is in an ethical vacuum.

The fact that some patients are actually not asked for permission to donate organ transplants has come under particular fire.

In 1996, the German parliament just barely prevented people from having their organs removed immediately upon death if they never gave explicit permission to do so. The wife of the German president set a public example when she publicly filled out an organ donor ID. Whoever wants to save or improve another person's life by donating an organ should do so, for there are far too few donors. Even today, most people do not think that the organs of someone who has just died belong to the public for the public good. Politicians and physicians must take this into consideration, even if in Germany alone two thousand people a year are hoping for a new kidney, five hundred for a new heart, and six hundred for a new liver, while another ten thousand Germans are on an apparently interminable waiting list for kidney donations.

The whole idea that organs could be removed from German citizens so long as they had made no explicit objection to being a donor during their lifetimes is decidedly premature. If, one day, most Germans wanted to volunteer their organs and the objectors were in the minority, then the "rejection rule" (i.e., that removal of organs is allowed except in cases where it is explicitly forbidden by the person in question during his or her lifetime) could be used. Until then, however, only the explicit agreement of a potential donor makes organ removal a possibility. The bizarre and sometimes shattering consequences of this just have to be accepted. With that in mind, the Three Star Japan company from Osaka offers private patients kidney transplants from the Phillipines—at half the normal price. The patient pays $500 to be placed on the waiting list and a fee of $100 for each month on the list. The total sum is then counted toward the final cost of the operation.

## CYBORGS, ANDROIDS, ROBOTS, AND ARTIFICIAL ORGANS

It is not so big a step from organ transplants to the concept of artificial human beings. Could they be created? They might be far more resistant to an inhospitable environment than real people. Actual robots would not be biological clones but would be pieced together by (replaceable) parts. These individual parts could be made of ceramic, metal, synthetic material, or silicon, or even real organs such as donated kidneys and hearts. The problems that arise with the reproduction of humans with the help of DNA or isolated pieces of tissue would be avoided here.

Of course, the joints wear out most easily. Today we have excellent replacements for knee, shoulder, finger, elbow, and hip joints. Other entirely different products could be added to that. From synthetic materials we could make eyeballs, nasal bones, tracheas, and veins. We could insert electronic parts as well. Based on our experience with pacemakers, there are now efforts to regulate urination with a bladder stimulator. A blood pressure regulator implanted in the shoulder could provide feedback to the brain through an electrical connection.

In 1995, research teams from various countries announced that they had developed an artificial retina for the eye. This retina is slightly similar to a movie screen—an image would be thrown onto this retina by the pupil and the lens of the eye. In people and animals, an image's path to the brain begins at the retina. The retina converts the image into weak electrical charges. The retinal image is not what counts at the brain's visual cortex; it needs electricity.

Scientists from the Massachusetts Institute of Technology, Harvard Medical School, Johns Hopkins University, and other research universities were able to develop a prosthetic retina as thin as a layer of cosmetic tissue paper. This image receiver has already been transplanted into rabbit eyes as well as three blind people. The results are promising, even if the eye researcher Ronald Bude from Albert Einstein College of Medicine in New York claims that it will take at least another fifty years before the artificial retina really functions properly. After 2000, the technicians from MIT are undertaking their experiments on human subjects only: people alone can judge how well the artificial conversion of images into electrical current succeeds. Dogs and cats are of little use in the initial phases of the experiment, since they cannot convey what they see. John Wyatt of MIT is convinced that a black cross on an artificial retina can already be recognized as such by the brain—an enormous achievement, unthinkable only a few years ago.

The long-range goal is an unobtrusive camera installed in the frames of glasses that would capture an image of the surrounding environment. The images would be converted into digital, "computer-digestible" information and then beamed onto the artificial retina. The prosthetic retina would convert the image into electrical currents and would send these currents to the nerves of sight. These nerves often still function in blind people, thus allowing their brains to be connected without great difficulty. This could even succeed after a brain transplant into a robot body.

The figure of the cyborg that appears again and again in novels is usually a creature with a robot body and a human brain: a brain in a functioning machine. This body could be compared to a real human body in many important respects. The question remains whether and if this will ever come to fruition. Still, when we read the futuristic notions of writers in past decades—completely fantastic in their own time—we realize how quickly even the most seemingly crazy technical advances become realities.

In a less pleasant version, the "human" robot consists only of a brain that feels and sees but is otherwise cut off from its environment.

Is it at all thinkable that a living brain could be implanted in a machine body? The purely biological problems could most likely be overcome. It is generally possible to create nerve connections to and from the brain sensibly and purposefully. Even the direct connection between humans and computers depicted in space science fiction novels could become reality in this way—maybe. There is work being done already on so-called brain chips, with which parts of the brain are supposed to be stimulated. Tiny sound receivers, implanted in deaf people's ears, will soon be directly connected with the brain as a sort of "neuroprosthesis." In principle, a robot consists of nothing more than a huge collection of interconnected brain chips and neuroprostheses. Whether the brain would be able to orient itself in a machine body is a different question entirely—it would probably be much more difficult than in a foreign biological organism.

The initial idea of the cyborg resulted in actual experiments. Robert White of Case Western University kept a monkey brain alive apart from its body for three days—during this time, electrical brain currents were able to be recorded, according to White. These signals supposedly were quite similar to if not entirely the same as those of a rationally functioning brain.

The future certainly holds several surprises for us in this area of research . . .

As this chapter has shown, neither technology nor morality ever seriously limited the human struggle with death. Our will to do what is possible is greater than ever, and the ethical reservations of the scientific community regarding the application of gene technology to humans, animals, and plants have melted away at the turn of the millennium like snowflakes in spring. More and more frequently, voices are heard— mostly in the United States—to make a case for a total lifting of limits where the cloning of human DNA is concerned. Already in early December 1997, only a few months after the huge controversy surrounding Dolly and Polly, the *New York Times* ran the two-page headline "Cloning Humans: From 'Never' to 'Why Not?' "

This was inevitable. Each of the techniques we have discussed here, from heart transplants to medical intervention after brain death, the resulting organ transplants, and finally the more recent advances in gene therapy, caused similar storms of protest. But as quickly as the objections flared up, they faded away, first among scientists and then among the general population. This is actually not especially surprising. For if it is a matter of lengthening one's own (valuable) life with new technology, who would want to turn down this possibility? No human being will want to resist the temptations of immortality in the future.

# CHAPTER 4

## HUMANITY—IMMORTAL?

---

*Life is irreversible.*

*NILS ARVID BRINGÉUS*

### WE'RE ALIVE ONLY IF THE EARTH IS TOO

What good is a long, healthy life if the earth we inhabit can no longer offer us safe habitat? Do not our current environmental practices suggest that the dream of eternal life will remain just a dream—if not for individual humans, then for all humanity? What are the chances for eternity, given the prospects of overpopulation, climatic catastrophes, and the extinction of species?

From a biological perspective, human life is an inconsequential, if active, part of the earth. Even without us—according to the scientist James Lovelock in 1979—the earth behaves like a "huge living organism. Like a living creature, it regulates and stabilizes the composition of the atmosphere and the climate so that they are optimal for life and for survival."[1]

Lovelock named this living organism after the Greek earth goddess Gaia. Some researchers do not like the idea of equating the supposed inanimate earth with a living creature, but the work of Lovelock's followers has confirmed again and again that the earth behaves indeed very much like a living creature. According to Lovelock, the earth adjusts its own environmental conditions so that life is possible. The salt content of the oceans, for example, always stays at about 3–4 percent, thus corresponding exactly with the demands of sea life. If the salt concentration were to exceed 6 percent, most marine life would die.

The air we breathe normally contains about 21 percent oxygen. Flying animals could not live with less. The precise measure of oxygen allows for maximum oxygen uptake to support a high-energy existence, as is the case for birds. Conversely, if the air consisted of more than 21 percent oxygen, every forest fire caused by lightening would inevitably be catastrophic. Just as one kindles a grill fire by blowing air on it, so the increased oxygen would accelerate the spread of fires that occur time and again (by natural causes) during seasonably dry summers. Under such conditions, even succulent plants would burn more easily. A single fire could devastate enormous areas of land or even parts of the earth.

As the earth has shown throughout its history, due to its natural balancing act, it is capable of surviving considerable environmental catastrophes. Humans play practically no role in this. They are merely a small part of a wonderful life network on which we depend—a very one-sided sort of dependency!

Stephen Harding, one of Lovelock's followers, who works at the International Environmental Learning Center in Dartington, England, fears that humans are not wise enough to draw the right conclusions from this. According to Lovelock, through the "3 C's," cars, cattle, and chain saws, we have gotten ourselves into a state that has caused many of our modern-day environmental problems. These represent the three most important reasons that earth's atmosphere seems to be growing warmer and warmer. The biggest lies with the increasing production of carbon dioxide as well as methane. Cars produce carbon dioxide by burning fuel. Chain saws cut down the forests that convert carbon dioxide back into oxygen. And cattle graze on cleared off fields and produce large amounts of intestinal gas (methane).

"Our influences," says Harding, "have led to a condition where the self-regulation we need can no longer be stabilized. We will most likely experience a jump in temperature over the course of the next 20–30 years to a higher level, and at that time, Gaia will try to stabilize herself again."

We know that people today can only adjust to gradual changes in the environment. A radical change in temperature could be devastating. For environmental scientists like Lovelock and Harding, there is only one way to use the earth's power of self-preservation: "Rethink things radically."

---

### The Psychology of Environmental Protection

Environmental protection is modern. It's fun. There are even excellent children's books on the environment. Today, hundreds of young people practice environmental protection in Greenpeace-Greenteams. Thousands of people are interested in plans for a progressive and simultaneously conservationist change in scientific practice (an example is the brilliant report to the Club of Rome, *Factor Four,* by Ernst Ulrich von Weizsäcker and the work of Amory and Hunter Lovins, as well as *The MIPS Concept* by Friedrich Schmidt-Bleek).

Some scientists are amazed, however, that environmental protection has shifted, with such force, into the public consciousness. Perhaps this could be explained with a simple idea.

At the head of the list is our natural wish for comfort. Vacuum cleaners, airplanes, frozen food—no one wants to abstain from such things. Most of all, no one wants to give up cars. Out of all Germans, 70 percent either never take a train or do so only once a year. Today, the car is the greatest essential desire and necessity in modern society.

Peter Marsh and Peter Collett, both psychology professors at Oxford, have wondered why people are so crazy about cars. Their answer: with cars, people can enlarge their sphere of influence, their "territory." They exercise power over other inhabitants within the narrow environment of motorists by driving aggressively, behavior that would otherwise be frowned upon in our society. A major European car manufacturer demonstrated this clearly in an advertisement: a car comes shooting out of a pistol, the caption underneath reading, "The trigger rests under your right foot."

For people, cars are an expression of themselves. A close personal tie with one's metal partner is hardly unusual. They are often decorated with gems, foxtails, and spoilers. In the United States, thousands of car drivers decorated their cars with plastic parts that ostensibly ward off insects. "For car drivers it absolutely does not matter that they are attaching useless, plastic paraphernalia to their cars, for they have found a way to give their cars a personal touch," psychologists March and Collett observe.

We humans are trapped in this "Catch-22": we love our creature comforts more than anything and won't relinquish them. Some won't give up their cars, others won't give up traveling to exotic places and buying exotic fruits or other products that could indirectly lead to damage of environments and habitats. Yet,

*(continued)*

at the same time, we know that the high energy consumption linked with these comforts and luxuries destroys our earth.

We are somewhat aware of the fact that we ourselves are responsible for extensive environmental destruction. But since no one wants to (or can) carry sole responsibility for the waste of energy and raw materials, the blame has to be given to someone else. To whom? Even "the government" is an insufficient scapegoat. Here's an example: we discover that jungles are being destroyed in South America. Wonderful! Immediately, articles are published, activist groups formed, and plots of jungle land are bought. Money spent in such actions doesn't necessarily bring us happiness, but it allows us to ease our conscience. Every bit of saved jungle represents a bit of our saved souls. At whose cost the money was earned in the first place does not matter.

Industrial acids stream into the oceans. Hurray! Demonstrations, flyers, and protest letters to soulless companies follow. Finally something to balance out the varnish and oil that we pour into our local drinking water.

An oil rig is set to be sunk into the ocean. English scientists assure us that the precipitation of suspended particles will completely bury the metal structure, including several drums containing poisonous waste, within 250 years. The researchers are convinced that the sunken oil rig will offer all sea creatures a quiet little corner until the waste has been assimilated underwater, underground. They explain, moreover, that we will have to sink our garbage deep into the ocean in a few decades anyway, since there won't be enough space on land, and storage will be too costly. But all the talking doesn't help. To unprecedented surprise, Greenpeace forced Shell Oil and the British government to give up. The wreck of the drilling rig *Brent Spar*—by now almost as famous in Germany as the power plant Chernobyl—is being taken apart on land. Why? "It's a matter of principle."

As useful as Westerners' activism for the rain forests and against oil rigs is, the enthusiasm for the environment in distant places could be connected with a psychological need for balance. We try and compensate for what we cannot or will not do in our own countries with what we do abroad.

## UNEXPECTED APPLAUSE

Many philosophers and biologists consider Gaia a beautiful fantasy. Its popularity among nonscientific people is explained by the Russian limnologist, Alexei Ghilarov, as resulting mostly from the memorable, catchy name. Yet the Gaia (often appearing as *Gaïa*) scientists find approval from another corner as well, from physicists and mathematicians occupied with the theory of complexity. In that abstract, lifeless realm of numbers, they have discovered mathematical formulas that when fed into a computer, create beautiful images. Some of these images look like ferns and are thus quite reminiscent of the living world. If these images, called fractals, are viewed under a strong magnifying glass (or if one carries out further mathematical calculations), we can tell that the original image consists of several smaller images that are identical to the larger picture. The scientists speak of the self-referentiality of the structures and show how these structures parallel the real biological world. Again and again, organisms produce self-referential structures, as in a fully grown tree whose branches seem to be smaller and smaller iterations of itself. If we observe plants closely, as in the leaf surface of a yarrow tree, we can trace images that closely resemble those on the computer—or vice versa. Physicists and mathematicians who "grow" such images in their computers think it possible that unknown physical regularities direct this phenomenon. "The deeper we probe into research," explains environmental expert Harding, "the more amazing the phenomena are, so that we are tempted to see Gaia as no longer a metaphor." The well-known mathematical physicist and author Paul Davies from the University of Adelaide in Australia and prolific British scientist and author John Gribbin take one giant step further. They claim that "entire galaxies" can be inferred from patterns of interconnection among organisms on earth. In the end, "an interdependent, living cosmic network" could take the shape of fractals examined here on this planet. A living universe? Would that be possible? For the moment, let's remain on earth and consider more closely the influences just mentioned that could destroy the present balance necessary for our survival on Gaia.

## THE WORLD IS BURSTING AT THE SEAMS

The thought that the earth is being choked slowly but surely by the masses of people living on it is nothing new. Hermann Römpp, a famous chemist in postwar Germany, described a phenomenon that today we insensitively call overpopulation: "It appears highly dubious whether man . . . will still exist in 100 million years. Today we may receive the opposite impression; human life is multiplying at such a rapid rate that, in a few decades, the earth will be filled to its capacity and will enter by natural necessity an age of mass hunger and poverty—if humans don't see the light."[2] These words are from 1948, and we haven't seen the light. While the rate of population growth is decreasing in modern Western societies, the number of people in economically disadvantaged but emerging countries continues to grow. The United Nations' most recent population estimate for the year 2050 lies somewhere between 9 to 15 billion, although women will give birth to only two children on average. The decline in the number of children born—in 1985, women gave birth to 4 children on average, and in 1996 to 3—diminishes the overall problem only a little. Only when every woman raises not more than 2 children will the world's population stabilize.[3] In the "Global 2000 Report to the President" (Council of Environmental Quality and U.S. Ministry of the Exterior, Gerald O. Barney, study director, Washington, D.C., Government Printing Office, 1980), we read the well-known estimation that traditional energy sources such as coal, oil, and natural gas will last only another 50–110 years if the population continues to grow at its current, rapid rate. Since continued increases in population growth seem inevitable, many experts anticipate a catastrophe. When working on this problem, biologists automatically, instinctively liken the conditions to those of the animal kingdom: in all cases, mass reproduction leads ultimately to the sudden eradication of many group members. The resulting smaller group can then begin to grow again until another collapse ensues.

## CHANGES IN CLIMATE AND THE CONSEQUENCES

Individual peoples or cultures could stand in as examples of individual species or populations in the animal world. Although many great cultures of humanity at some point vanished, humanity as a whole has thus far survived. But at several critical points in the history of the earth, nearly complete extinctions of all life occurred. Most paleontologists and zoologists agree that the earth was populated by life forms several hundred million years ago, the likes of which do not exist anywhere, as evolutionary descendants or otherwise, today. These life forms disappeared about 570 million years ago, leaving nothing but their fossilized remains. Then, the Cambrian explosion occurred, causing the rapid unfolding of a richly diverse animal and plant world. No one can say for certain what caused the mass extinction of those earlier, primitive forms of life. We know just as little about what put an end to the Mayas, Egyptians, and Romans, even if there are more than a few interesting speculations on the matter. For example, changes in climate and resultant widespread famine are possible explanations. And it is precisely catastrophes such as these that scientists warn us about today. Apparently, however, there is a serious difference between then and now. A new climate catastrophe this time could rest with humanity.

No one can say for sure what specific changes will take place in our environment in the future. Only one thing seems clear: global warming is happening. Since 1850, the average temperature on the surface of the earth has risen 1° F (0.5° C). Of course, this rise could be due to normal, smaller changes in the climate (along with greater though less frequent changes) that have always taken place. The conclusion of human and technological responsibility for global warming would then be thrown into question.

Professor Christian Pfister of Bern, Switzerland has been occupied with this issue and is looking for weather reports more than 150 years old. During his investigations, he has had to convert old units of measurement, decode calendars, and translate notes such as "great snow" or "much and strong wind" so that these expressions could be catalogued into decipherable tables and charts.

Pfister's results: since 1985, the only other time world temperatures were as warm occurred in the years between 1530 and 1564. We are obviously in a similar period of global warming. But, he adds, "the overview of the weather and climate history of the last centuries demonstrates that the natural range of variation and fluctuation in our climate is much greater than had been assumed for a long time." He considers our recent "springlike, snowless winters" to be especially meaningful. According to Pfister, if our winters continue to have little snow, we can assume that humans really are responsible for current changes in the climate. This assumption has found further support in the 1997 results of a team of thirteen American, English, and Australian researchers under the supervision of climate expert B. D. Santer from the Lawrence Livermore National Laboratory. The scientists linked the existing measurement results for the ozone, sulfuric oxide, and carbon dioxide in the air and the stratosphere with the rise in temperature between 1963 and 1987. "Our computer model increasingly matches the actual data," the researchers report. This means that the assumption calculated in the model—that is, that industrial and household gases have an influence on the climate—was confirmed. "While it is indeed probable that human behavior is at the bottom of this result, there still remain incalculable elements, particularly as regards naturally occurring environmental changes."

The majority of climate experts, however, are far less timid in their judgment than these scientists and unambiguously affirm human responsibility for the changes in the earth's surface and atmosphere. They are convinced that we will only lose time if we continue to wait around for additional studies. The possible consequence: it will be too late. There is one rather unsatisfactory consolation: if people are not the cause of global warming, then we will probably not be able to stop it.

One consequence of global warming could be a rise in sea level and the subsequent submersion of vast parts of our continents. Scientists have already observed that regions previously covered by ice are now exposed. John Fowbert and Lewis Smith, two British South Pole explorers, discovered that the grass *Deschapsia antarctica* has spread out twenty-five times its original area within thirty years. At the same time, the temperature rose about 4° F (2° C) in the summer—a huge difference for meteorologists.

Some people in the Western world might welcome the idea of warmer pastures, more sun, palm trees, and parrots, but in the face of the catastrophic consequences global warming could bring, such day dreams are naive. Hence the headline from the journal *Science* in February 1995: "When the temperature rises, so do dangers to our health." What is implied here were not the familiar life-threatening problems from heat, like blood circulation problems, but diseases like malaria, dengue fever, and sleeping sickness. The carriers of these illnesses—certain insects and other organisms—could multiply in higher temperatures worldwide. Several researchers have already calculated how high the risk of infection for such diseases would be in certain regions of the earth by the year 2055. It is a terrifying picture. For example, Europe, Africa, South America, and the United States will be nearly completely and continuously threatened by malaria.

There is still no effective immunization against malaria. And mosquitos carrying the often lethal pathogen represent but one small part of a broad range of likely catastrophic consequences of global warming.

Today an attentive newspaper reader is aware of the connection be-tween rising temperatures and the increasing amount of exhaust fumes from industry, traffic, and households. The main culprits are industrial-ized countries that for years have been discussing the subject in environ-mental summit conferences without coming to any definitive guidelines to stop these developments. Ironically, they try to dictate environmental protection regulations to emerging, developing countries whose tenuous economies make implementation of such regulations all but impossible. Today it is abundantly clear that a general amelioration of living stan-dards in poorer countries could merely intensify current environmental problems to unrealized proportions.

In the future, more carbon dioxide will undoubtedly be emitted into the air and will spread itself throughout the earth's atmosphere. This mantle of air, now altered by carbon dioxide, traps the sun's heat rays and no longer releases them back into space. A similar process occurs under the glass panes of a greenhouse, which is why the current change in atmosphere is referred to as the "greenhouse effect." But this is only a rather cursory, roughly sketched description of a far more complex phe-nomenon, influenced by indirect effects and feedback.

In principle, the atmosphere's absorption of a little heat is a useful thing. Otherwise, the surface temperature of the earth would be -18° C—too uncomfortable for us humans and too cold for many plants and ani-mals. What consequences would increasing heat have for flora and fauna?

## WHY DO WE NEED A DIVERSITY OF PLANT AND ANIMAL SPECIES?

A worldwide rise in temperature would cause some animal species to multiply and others to die out. But doesn't the disappearance of some species balance itself out through the propagation of others?

There are two answers to this. First, there are more animal species dy-ing today than ever before.[4] Second, the death of even a single animal species can have terrible consequences, since many organisms are mutu-ally dependent within habitats.

Charles Darwin considered this very modern problem. What kind of mutual relationships exist between organisms of differing species, he wondered, and what would be the connection with his ideas on the "survival of the best adapted." Today, we must pose the question in reverse: what would happen if organisms disappeared from the food web? Here is an example from Darwin himself, summarized by French author Yves Delage with M. Goldsmith:

> Certain insects are necessary as pollen carriers for the pollination of several flowers. Such visits are indispensable for several kinds of clover plants—red clover, in particular, requires the bumblebee. What would happen, then, if bumblebees became rare in England, or even disappeared entirely? Red clover would itself become a rarity or even disappear entirely. The number of bumblebees largely depends on the number of field mice, who destroy their honeycombs and nests. The mouse population is dependent on the number of cats. In this way, the fertility of red clover is determined by cats![5]

While Darwin thought about his issue only theoretically, the U.S. army tested it unwittingly, though no less inexcusably, in a real experiment. This experiment was not with red clover and cats but with trees and insects in Vietnam. Agent Orange, the highly toxic defoliant, transformed enormous, dense forests and swamps into steppes. The effect desired by the military caused fundamental changes in many ecosystems of Vietnam. The American zoologists Gordon Orians and E. W. Pfeiffer reported on the damages to the environment when they were in Vietnam as early as 1969, four years after the war had started, with the support of the Vietnamese authorities and the Society for Social Responsibility in Science. Over enormous areas of the Mekong river delta, typical "swamp trees," in particular the mangroves (*Rhizophora*) with their bizarre roots and branches, were completely stripped of their leaves.

Trees especially sensitive to the effects of Agent Orange did not recover for decades. "We cannot exclude the possibility that the original forests will not regrow," the researchers reported laconically. Forests of breadfruit trees (*Artocarpus*), an important local food staple, and several of American rubber trees (*Hevea brasiliensis*) were unintentionally destroyed: the wind had scattered the poison. The forests' devastation caused the collapse of several food chains. "On our way through the defoliated forests, we saw not a single insect- or fruit-eating bird with the exception of swallows (*Hirundo rustica*) that had flown from the north. . . . The fish-eating birds didn't seem to have it so bad, even if there were many fewer of these birds than we had expected." The zoologists tallied no more than sixty-four individual fish-eating birds during a two-hour boat ride through what had previously been a forest heavily populated by birds. It was now little more than a dead zone. "The only other vertebrate we saw was a large crocodile (*Crocodylus*)," reported Pfeiffer and Orians. From plants to insects, fish, and birds, to large vertebrates, the forest inhabitants were dead. A defoliant once considered safe for use around animals had liquidated fundamental food sources for many animal species,[6] which of course had an impact on food sources for humans. "Although hunting was forbidden because of the war, armed people were moving through the forests, most of them undernourished. They apparently shot all available animals," the scientists observed, and yet: "The tiger alone gained an advantage from the war. Over the last 24 years, they have learned to associate gunshots with the dead and wounded. Tigers would come running at the sound of gunfire and ate the victims of war." But the hunters could also turn quite suddenly from the attackers to the attacked. The war claimed human lives just as it did animal lives. Those who did not die were in danger of being eaten by tigers. What choice did people have but to flee to the cities? "Within the last 10 years, Saigon has gone from being a quiet little city with 250,000 inhabitants to an overpopulated city with 3 million," say Orians and Pfeiffer. "Moreover, the enormous American accumulation of capital [in Saigon] led to more and more motor vehicles on the streets. . . . There were constant traffic accidents. The air pollution from a combination of fuel and oil has become so bad that many of the trees along the main streets have died or are dying." Hunger, stripped forests, and a boom of Japanese motor vehicles in Vietnam—this was the sad epilogue to the wasting of many dollars and a lot of defoliant.

We don't need a war, however, to cause species to go extinct. We also don't need to travel as far as Vietnam to observe the extinction of species: a visit to the North or Baltic Sea can drive biologists to despair.

## OVERFISHED WATERS

In the North and Baltic Seas, the demise of the herring is apparently a decided issue. While the European Union prescribes economically reasonable and politically defensible fishing quotas in all its member countries, these are continually ignored. In the past few years, fishermen from the Netherlands are the main culprits of overfishing for herring, at other times, the guilty parties hail from elsewhere.

Worse than ignoring quotas are fishermen who engage in overly aggressive fishing practices. Faced with unemployment, many fishing nets from Denmark and the Netherlands harvest their catch illegally. The workers cannot be accused of not caring about protecting the fish. They often have no idea that they are depriving themselves of a basic food source. (In 1992, I myself saw how fishermen from western Ireland refused to return live squid or octopus captured during fishing to the sea, since they have long believed that these animals are vicious predators that would destroy their catch. If by chance they hauled in a squid or octopus, they killed it. Even when they were offered a financial bonus if they placed a live squid in a water-filled container, they complied only with great reluctance.)

Nevertheless, today the number of North Sea herrings is one-third less than the amount necessary for the natural preservation of the species.

This is a new record low in the herring drama, which began in 1967. Over twenty years ago, Norwegian fishermen thought the herrings had gone mad, because the enormous schools of fish did not take their route over the North Atlantic to the coast of Iceland, where they usually spend the winter, but simply moved in the direction of Spitzbergen in the Barents Sea.

The fishermen could not accept this attempt to escape from their nets. With a new, ultramodern fishing fleet, they followed the herring and made the best catch ever, given the conditions: 2 million tons. The fishing technique of localization that had been introduced in 1962 for large schools of fish had yielded tremendous success. But it did not take long before catastrophe struck. Far too many herring had been removed from the water. In 1977, the European Community called for a total stop to catching fish between June and October, a sanction that had to be extended through 1982.

As German zoologist Vitus Dröscher reports, the Commission of the European Community was influenced in its decision by the example of the Norwegian Lofoten Islands. In 1969, not a single herring could be caught there—the waters had been fished clean. All studies indicated that regeneration of the herring population seemed unlikely. This brought the local fishing industry to near ruin, and the sea birds began to die in droves. The death of these birds was a terrible experience for environmental activists because, among other reasons, protectors of the environment tend to be great bird lovers.[7]

At present, the concept of the extinction of species does not appear especially catastrophic. Again and again we see signs that, even under the worst conditions, animals and plants are not that easy to kill off. Personally, I have discovered numerous ground beetles, which are specially protected by German federal law, as well as twelve different kinds of dragonflies on an expanse of land used by the building construction industry.

People do not usually live long enough as individuals to perceive the real extent of species extinction with their own eyes. Many animal species die out very slowly. Extinction is most apparent when anthropogenic forces create desertified regions, such as the ones we create on ocean floors.

## INTENSIVE FARMING

The variety of species so essential for our survival is also endangered by modern farming methods, the development of which seemed inevitable in order to feed a growing world population.

In 1798, the English economist Thomas Malthus determined that the rate of human reproduction greatly exceeded the rate of food production. As a result, Malthus urged birth control—in the form of abstinence. But he was realistic enough not to expect widespread compliance with such a demand. He believed that, in the end, only epidemics and wars could keep the number of people on the planet low enough to ensure sufficient food for all.

In industrial countries, however, we have succeeded in producing more food than we would have dared to hope. The fact that apples, cauliflower, and tomatoes in Europe are dumped by the truckload into the oceans to avoid the collapse of market prices from overproduction is no secret. In other parts of the world, the situation isn't so sweet. Most people feed themselves almost exclusively on rice, wheat, and corn. An example from the Philippines: throughout almost the entire country, each meal, including breakfast, consists of one or two cups of cooked rice and a small side dish—vegetables or a piece of meat that Westerners would consider tiny. Each year, 3.5 million people on earth die of starvation because they cannot afford even a modest meal.

Would it be desirable, then, to introduce our technological innovations along with the planting of crops in poor countries? This question cannot be answered with a definitive yes, since the intensive farming methods practiced over several years also have severe disadvantages that have recently gotten clearer. First, the overfertilization of the soil and the use of highly noxious pesticides have overwhelmed our land, the groundwater, and the surface waters. Second—and this is the more lasting damage in the end—we reduce our future chances of survival by concentrating on fewer and fewer species of useful plants and animals, by cultivating plants as monocultures.

A simple example: at the beginning of the century, there was still a tremendous variety of different kinds of apples. Today, only a select few are marketed. The problem is not that we must deny ourselves of as many different tastes of apples or other fruits as possible. If we content ourselves with growing only a few kinds of useful plants and breeding only a few kinds of useful animals (no more than about one hundred species are necessary to feed and sustain humans), we limit the possibilities of adaptation to changes in the environment. As just explained, a variety of species and variations within a species are necessary to ensure survival. Artificial forests that contain only one kind of tree highlighted the problem of imbalance in the 1970s: a single sort of pest could destroy entire stretches of woodland. This would not have happened if many different types of trees had been standing together, because many destructive animals show a preference for only one kind of plant. And what goes for forests goes for useful plants all over the world. Moreover, monocultures exhaust the soil, thus increasing the necessity for fertilizer and other artificial means to keep crops producing.

If the earth's population continues to grow, it will become necessary to alter our farming methods. There are already options for solutions, all of of which may be possible, thanks to consumers who have demanded such through their shopping habits and demonstrated desire for perfect, unadulterated foods. More and more farmers are considering organic farming and methods of species-friendly breeding. And scientists are not standing idle either. Many of them—in universities and industries—

support responsible action in the huge project of ensuring basic food sources for the world's population. This includes the cultivation of especially resilient kinds of grain. The kinds of grain found today withstand wind and harsh weather only if they are cared for continuously, and they are dependent on very specific climate conditions. In many regions of the world where the climate is more difficult, these kinds of grain cannot be cultivated. In order to cultivate new kinds, researchers are going back to nature's genetic treasure chest. They use the characteristics of wild sorts of grain that have persevered during the course of millions of years in nature's natural selection process.

Yet the prerequisite for the use of such species today is and will be in the future the protection of their genetic codes. Seed banks for plants have been set up in several countries for this purpose.

---

### Herbaria and Wild Tomatoes

Every type of plant originates from a long chain of unrepeatable coincidences. For this reason, the genetic information found in a plant and in all other species is in a certain sense more than merely the ordering of letters of a genetic code. A biological species is the singular result of a particularly good adaptation to current and earlier life conditions. Since more species die out today than ever before, several organizations and institutes are in place to at least save the genetic information contained in the disappearing species (figure 13).

In Germany, the gene bank of the Institute for Plant Genetics and Cultivated Plant Research (IPK) in the town of Gatersleben and the Federal Institute for Cultivation Research on Cultivated Plants (BAZ) in Braunschweig have taken over the job of collecting all living plant species. Of the hundreds of thousands of plants to be stored, the institute in Gatersleben has already collected several hundred. Other research institutions collect every kind plant useful for human nutrition.

The Consulting Group on International Agricultural Research (CGIAR), founded in 1971 by the United Nations and the World Bank, mediates among all the large agricultural institutions that try to save crops over time. In 1988, the CGIAR seed banks received more than 350,000 kinds of useful plants in dormant stage, among them almost 80,000 kinds of rice and 30,000 kinds of beans.

An issue at hand is where to store such vast repositories of seeds and plants. The plant seeds cannot be stored in space, even if space agencies such as NASA

*(continued)*

have considered the possibility. Several manned space ships have shown that the seeds lose their potential to germinate dramatically after only a few weeks. Whether cosmic radiation is responsible for the destruction of the seeds remains to be clarified.

Many botanical gardens and university institutes have been discussing the possibility of storage in herbaria for a long time. These used to serve as "comparative collections," that is, newfound plants could (and still can) be compared with precisely examined, stored samples.

More than leaves, twigs, and blossoms are stored in herbaria; their seeds are kept as well (figure 13). Since seeds stored under dry and dark conditions can germinate after decades—even after centuries—there is the hope that their genetic makeup will be archived together with the seeds.

But this is an almost futile race against time, since it is impossible to collect and store all plant forms, with all their different variations. In this respect, the futile collection of seeds or DNA is no more valuable than any other museum collection. With a lot of luck, however, a technique might one day be found to reproduce original, once living plants. It remains questionable what meaning this would have in a future, different environment.

Seeds from herbaria have already proven to have real agricultural use in several cases. A few years ago, for example, a new "improved" kind of tomato was developed. It originated from the cross of a common type of tomato with a wild tomato cultivated from herbarium seeds. The wild seeds came from an expedition to the Andes led by researchers of the Royal Botanical Gardens, Kew (UK) in 1962. The late result of the expedition: a more nutritious fruit for the consumer and annual increase in turnover of 8 million dollars for the U.S. tomato industry.

That the protection of species can pay off was determined by the German Federal Ministry of the Environment in 1994: every U.S. dollar spent on species protection programs brings a profit of two to three dollars.

## EMIGRATION INTO SPACE?

Death is a fact of life. Today we know that civilizations die as well. But the possibility that the existence of humanity as a whole might be threatened by our unreasonable and wasteful lifestyle seems new to us. Some scientists are convinced that the problems described here will spin out of

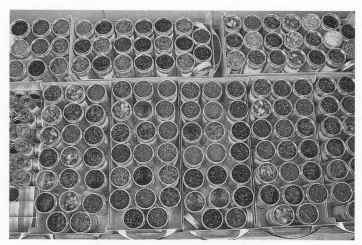

13 Plant seed canisters in the Millenium Seed Bank of Kew. Throughout the world, researchers are attempting to collect the seeds and DNA of plants. The largest collection to date is in the Botanic Garden of Kew near London. British scientists set the year 2000 as their goal to collect, in their seed form, all fourteen hundred living plants in Scotland, Wales, and England for the Millenium Seed Bank. They even hope to have twenty-five thousand types of plants in the seed bank by the year 2010—a good tenth of all plant types on earth. (Photo: Royal Botanic Gardens in Kew)

control and that emigration to other planets would be a way to escape this dilemma. Of course, we must wonder whether space researchers promote this thesis to get research grants for their expensive projects, since scientific interest in the colonization of space goes back to the beginning of space travel itself. As early as 1952, five years before the first satellite (Sputnik 1; Russian for "companion 1") was launched into space on October 4, 1957, Wernher von Braun had published *The Mars Project*, in which he outlined concrete plans for a flight to the neighbor planet. Science fiction writers have always been fascinated with the notion of human settlements in outer space, even before scientists. Although I have enjoyed Isaac Asimov's stories, I had never really given any real credence to these ideas—until research for this book brought them back to my attention.

The point of departure was a meeting with the inhabitants of Biosphere 2 (the originators of this large experiment named it Biosphere 2, with the understanding that the moniker "Biosphere 1" belongs to

14 The first artificial living environment for humans and plants, the airtight, closed metal vat of the Russian scientist Yevgeni Shepelev. In 1963, the scientist at the Institute for Biomedical Problems in Moscow locked himself in this vat of 15 cubic feet, connected to a biolung (a container full of algae and water), under the supervision of his colleague Ganna Meleshko. The algae transformed the carbon dioxide Shepelev exhaled back into oxygen. The experiment was a success. (Photo: G. Meleshko)

Earth), an enormous glass house in the Arizona desert, where eight researchers lived and worked for two years to find out whether humans could survive in a sealed, self-sustaining biological system.

The first experiment of this kind had taken place in the Moscow Institute for Biomedical Problems in 1963. Back then, Yefgeni Schepelev locked himself under the supervision of his colleague Ganna Meleschko into a metal container of only fifteen cubic feet, which was connected with a biological lung (a container of algae in water). The algae were supposed to transform the carbon dioxide Yefgeni Schepelev exhaled back into oxygen. The experiment was a success—the brave researcher lasted a day in the container. However, he terminated the experiment with the words, "May I smoke one last cigarette?" (Schepelev is, in fact, still a chain smoker today and jokingly advised the Biospherians to use

15 The work and living block of Biosphere 2 in the Arizona desert. Eight researchers lived and worked for two years in this artificial world to find out whether humans could survive in a closed biosystem. Beneath a beautiful glass dome, there was a tropical forest, a sea, a steppe, fields, and a 27-foot waterfall. (Photo: Nancy Dise)

cigarette smoke as replacement oxygen in case their regular oxygen source expired.) The idea of an enclosed biosphere was developed by the Russian geochemist Vladimir Vernadski at the the beginning of the twentieth century. Yefgeni Schepelev and his colleague Josef Gitelson continued their work on it for decades afterward. In the end, a later experiment using a fifty-four-gallon tank of algae sufficed to provide a person with enough oxygen for any length of time (figure 14).

But despite extensive preliminary work, the Soviet Union never put together as expensive a project as John Allen, the original cofounder of Biosphere 2, together with the financial sponsorship by an American millionaire.

After six years of preparatory work, a closed off, self-contained world with fields, rain forests, a miniature ocean, a steppe, and a twenty-seven-foot waterfall arose under a translucent dome (figure 15).

Since the building structure of Biosphere 2 is in a region where irradiation and temperatures fluctuate dramatically, all the plants and animals in the enclosed system had to adjust to an entirely new environment: most of them came from the tropics, where there are hardly any seasonal climate fluctuations.

Animal breeding proved the least problematic. The Biospherians intentionally chose only a few hardy Nigerian dwarf goats and a specifically bred strain of resilient chickens for this great experiment. The six thousand cubic feet of fields yielded four-fifths of their entire food supply, including fruits such as papayas as well as grains. The rest of the food came from stored supplies of grains that would have been replaced by self-grown grains in a longer expedition; the residual food sources were inconsequential in the final balance.

The Biospherians ate about twenty-two hundred calories per day and lost about 15 percent of their body weight in the first six months. The fieldwork took up about a quarter of their work time, one-tenth of their time was devoted to the animals, and another tenth to food preparation.

The eight scientists' dream might have been realized in 1985, but there was one significant problem. Biosphere 2 was losing oxygen. After more than one year, the oxygen content of the air was just above 14 percent (the normal oxygen content of air is 21 percent); it was raised to 16 percent with the addition of oxygen from the outside. But this was not the only problem: the enormous machinery needed for water and air circulation, the ultraviolet radiation used for the purification of drinking water, the various technical appliances and ventilation devices needed for the cooling inside Biosphere (which, without cooling, would have reached a temperature of 302° F)—all this required enormous amounts of energy. While a manned space station can accumulate enough energy through solar cells, systems like that of the Biosphere 2 had to be secured two and three times, or else the inhabitants would die in the case of breakdowns within the system.

Even if in the future it were possible to create technically impeccable conditions and a perfect air and water circulation system, nature might still thwart our plans in the end, since it would obviously (still) remain unpredictable. A small example will demonstrate this: the scientific head of the Biosphere team, Abigail Alling, noticed unanticipated changes while examining the coral reefs that had been built within the structure. They had transplanted the coral from Mexico into the Biosphere's 750,000 gallon mini-ocean. During the course of the two-year experiment, some kinds of coral seemed to thrive while other colonies of coral dwindled. All corals lost about 25 percent of their tissue, while, at the

16 The eight bionauts. This photo shows the inhabitants of Biosphere 2 two months before their release from the artificial world. *From left to right:* Mark Nelson, Abigail Alling, Taber McCullem, Sally Silverstone, Mark van Thillo, Roy Walford, Jane Moynter. (Photo: Mark Nelson, Biosphere Ventures)

same time, eighty-seven new colonies formed. In comparison with a preliminary experiment, in which algae had grown over the coral, killing it, this was tremendous progress, but the exact conditions for maintaining coral remain unclear now as before.

Another difficulty that would have led to catastrophe over a longer period was the rapid rate of death of the insects and birds that had been brought into the biosphere to pollinate the flowers.

The Biosphere 2 project was not the only undertaking of its kind, but it was certainly the largest. Of course, NASA has been concerned for years with the question of whether and how survival would be possible in artificial environments. In this spirit, the aspiring astronaut Ray Collins is experimenting in Alaska with a one-man cell that could be built on Mars and would provide an astronaut with food up until his return to Earth—many, many years. Another small artificial world—"the most southern nursery in the world"—already sits on the grounds of the McMurdock Research Station at the South Pole in Antarctica. This station has operated since 1996 with the joint effort of NASA and researchers from Purdue University. The McMurdock Station is intended to assist in the vast research necessary for preparation of a moon station (planned for 2010) as well as a Mars station (planned for 2020).

At present, we have not made it very far. The NASA experts haven't solved scores of problems. One of the most serious ones is the burden on the astronauts' health as a result of little to no gravity. Who knows whether such problems can ever be solved. Even if all our technical questions are answered one day, there would be another problem: the human psyche. Who would want to leave our home planet for a vast period of time? When I asked the Biospherians if they would risk a flight to Mars, Mark Nelson, the initiator of the Biosphere project, answered, "With a return ticket?" I said, "Yes," at which point he decided without hesitation, "Then I would go—even for ten or twelve years!" Biospherians in particular are well aware that it isn't easy to live together with a limited number of people in an artificial world. Technical problems would be compounded by psychological ones. By the end of the Biosphere 2 project, eight friends had become eight colleagues (figure 16). The psychological problems that arise from such long-term projects should by no means be underestimated. There are even NASA researchers who, despite their great enthusiasm for space travel and willingness to take the long flight to Mars, openly admit that an astronaut's psychological hurdles might prove to be the greatest—perhaps an even insurmountable barrier. The Biosphere 2 example has shown us that we cannot avoid the unanswered questions that concern us on our home planet—overpopulation, changes in climate, the decimation of species, among others—by emigrating to other parts of the universe, not because we lack the technology, but because we cannot foresee the myriad complications that may hobble such explorations. We must devote ourselves to the solution of some of these difficult questions if we want to survive as a species on this earth.

## CHAPTER 5

# THE MEANING OF LIFE—
# "BIOLOGICALLY" SPEAKING

---

*Haydn was right. All creative forces are stronger than death.*

NAPOLEON TO JOSEPHINE AT THE CHRISTMAS PREMIERE
OF HAYDN'S CREATION, CITED BY E. W. HEINE

## THE DECODING OF HUMAN DNA

I've intentionally reserved one last prospect for the future for the conclusion of this book, inasmuch as it leads us closer, perhaps, to the secret of eternal life than all the other efforts previously described: the most comprehensive study of human DNA ever. Piece by piece, building block by building block, the complete composition of DNA is now illuminated as the human genome. And, somewhere in this overwhelming amount of information, the instructions responsible for aging and death are hidden.

Although there are differences in the human genome of each individual, its structure is by and large the same for all people. For example, the instructions for gender, eye color, etc., all have the same location on the DNA strand for each person. Since the genes at a single location differ only in minor ways between two people, the fact that they have the same function for each can be recognized immediately. We cannot be sure what kind of knowledge can be acquired or what sort of other use genetic decoding will yield. Geneticists' opinions range from "boring" (by which they mean the tiring, monotonous work of decoding with the help of machines, which does not help to understand anything by itself) to "fascinating" (by which the project's potential is implied). Its most obvious use value will clearly be in the medical world. The more biomedical scientists know about genetic substance and alteration from disease, the greater the chances are for a cure. Other insights won't make big headlines perhaps, but are in my estimation at least as interesting. First among such exciting questions is why more than nine-tenths of human genetic makeup consists of what was formerly known as "junk DNA," building blocks that do not make up genes. "Junk" DNA can, however, assist in matching biological traces such as blood, hair, sperm, etc. that can be attributed to a specific person (DNA typing or a "genetic fingerprint"). But we know next to nothing about its function in a living body. According to one hypothesis, "junk" DNA is indirectly involved in the building of the body: it can join itself together with loops, hooks, and other structures, thus influencing to what extent the working order is for a "useful" section of DNA. For example, when a "useful" area is shielded or protected by clustered "useless" areas, its information can no longer be accessed. Since not all genetic information is needed at all times, this might be a way to switch off useful but, at that moment, unnecessary DNA sections. Conversely, such clusters might switch on dormant genes. This switching is only one of many possibilities. We will probably soon encounter entirely new processes that today seem unimaginable.

The DNA strand of multicellular organisms exists at certain times in the form of many tightly coiled loops of DNA. These loops are called chromosomes. In 1992, a commission of the European Community reported the first complete decoding of baker's yeast (*Saccharomyces cerevisiae*)—thirty-five laboratories in seventeen countries had worked on the project. It took another four years until it was made official, on April 17, 1996, that all sixteen yeast chromosomes had finally been untangled. Primarily European laboratories were involved in the decoding, having worked under the supervision of the Belgian professor of chemistry André Goffeau: seventy-four of the seventy-nine participating laboratories were in Europe.

The huge masses of data uncovered through decoding are analyzed in laboratories specifically set up for that purpose. In the case of the yeast, the job was taken on by Werner Mewes's research team at the Max Planck Institute for Biochemistry in Martinsried. "This is a milestone in the history of biology," asserted Mewes, once the yeast DNA had finally been broken down into its various components. A similarly important step was the discovery of the DNA-structure of the single-celled pathogen *Haemophilus influenzae* (which, despite its name, has nothing to do with the flu). Very quickly, laboratories all over the world got complete DNA sequences of so-called model organisms like the laboratory mouse, the fruit fly *Drosophila,* the 1-millimeter-long nematode *Caenorhabditis elegans* (*C. elegans*), and the mustard weed *Arabidopsis.* The decoding of the genomes of those model organisms had been an initial attempt to explain how human genetic makeup could be studied quickly, reliably, and as automatically as possible. All the same, human DNA was expected to be about one thousand times as long as baker's yeast and about eighty times as long as the DNA of *C. elegans.* Yet, in February 2001 the human genome sequence was announced to be completed. Sequencing was performed with major contributions of a worldwide net of researchers and is undergoing further refinement. (An interesting development will be the involvement of private companies in helping push forward and raise public awareness of the international Human Genome Project that started before them. People like Craig Venter of Celera Inc. who, bypassing traditional technical and administrative hurdles, might contribute to establishing alternative routes for natural sciences research in the new millennium.) One major surprise was that a lower number of genes were found in human DNA, with earlier estimates ranging to over 140,000 genes. Because of the relatively low number of genes, it might be a good idea to look elsewhere for the mechanisms that generate the complexities inherent in human development and the maintenance of body functions. Best candidates are the noncoding stretches that make up more than 95 percent of our DNA and contain regulatory elements moderating gene activity.

Understanding the human genome was only possible because of work previously executed on biological "model organisms." Without having studied the genetic makeup of *C. elegans* (figure 2), the fruit fly, and the mouse, DNA-analysis of humans would be—at least medically speaking— fruitless. Why?

Next to no experiments are permitted on human beings. A human being lives a long time and does many things one cannot forbid, things, moreover, that would disrupt close scientific observation, such as smoking or not smoking, eating too much, eating too little, dancing or not dancing, and so on. The model organism *C. elegans* is the exact opposite.

It is small, translucent, lives only two weeks, can be bred on a layer of jelly with a single kind of bacterium for food, and has numerous offspring.

In addition to that, biologists John Sulston, at the Sanger Centre in Cambridge, Einhard Schierenberg, now at the University of Cologne, and their colleagues had already described the origin of all the worm's cells in the course of its growth back in 1983. Since then, the point and location at which a cell in the body of a *C. elegans* divides can be calculated almost down to the minute. And, what is most important: already a whole slew of *C. elegans* genes are known that deal directly with aging (see part 1).[1] If they were to be changed or even switched off entirely, the worm would live longer than its genetically normal fellow species members.

Since *C. elegans* has been investigated so thoroughly, the organism is the subject of even more specific, targeted experiments. One example: several of its DNA sequences begin with 3 consecutive DNA-building blocks, abbreviated with *C, A,* and *T* (CAT), and ending with *T, A,* and *T* (TAT). Similarly, when we find a human DNA-segment, we try to discover its purpose. To do so, the order of the DNA-building blocks are entered into a computer. The computer compares them with all other known DNA sequences (mostly genes) from all other animals that have been studied. Since all organisms are interrelated, to a great degree, very similar sequences are frequently discovered. This is commonly the first step in the search for the meaning of a human DNA sequence.

If comparable DNA sequences have been found in other animals, scientists then test (again with computers) whether the biological meaning of the sought DNA fragment is already known. This is often the case, at least at the level of metabolism.

For example, when a DNA building block order found in humans has something to do with aging in *C. elegans*, it might have something to do with aging in humans too. Granted, humans and *C. elegans* do not seem to be closely related to one another, yet, at the genomic level, it is surprising how similar all living creatures, especially higher animals, turn out to be. Naturally, it would be better if a human DNA fragment resembles that of a vertebrate, as in the South African clawed frog *Xenopus* or—even better—the rat. Among the most often used experimental model organisms, mice probably resemble humans to the greatest degree.

If we were very lucky, we would find a similar area of DNA in all three animals. The second best case would be knowledge of the DNA's biological function in the mouse. Then we could be almost certain that the DNA fragment had a very similar function in humans. If we found a comparable DNA fragment only in *C. elegans*, we could still study whether the worm developed differently if the DNA in question was altered in some way. Since *C. elegans* grows overnight and is largely transparent, changes that differ from the normal course of development can be seen immediately—for obvious reasons, we cannot and do not want to carry out such experiments on humans.

How ironic it is that even a spineless little worm can help us achieve a better understanding of human life. Without the help of model organisms such as *C. elegans*, the data from the decoding of human genes would be difficult to analyze. Recall the earlier discussion about another human gene linked with aging: it contains instructions for an enzyme called helicase that varies in patients with Werner syndrome, who age prematurely and look like elderly people in their early years (see part 3). In November 1997, Japanese scientists published their findings regarding a gene called *klotho*—named for the Greek goddess of fate, who spins the courses of all human lives. The scientists switched off the gene in mice; the result: the animals showed premature signs of aging.

Today, we turn to the words of James Watson, the codiscoverer of DNA structure. At first, Watson had actively supported the Human Genome Project, but resigned from his position leading it at the beginning of the nineties because he had a falling out with several scientists involved in the project. Still, he was acknowledged by the leaders of the human genome sequencing endeavors to be a source of inspiration.

The heading of the last segment of the book he wrote with three colleagues, *Recombinant DNA,* reads, "Humanity will profit from knowledge about our genome." I would like to cite this passage in full because it precisely captures the thoughts of current geneticists and biomedical scientists. It might come to pass that we think back on these words.

We believe unconditionally in the usefulness of new knowledge in genetics—in research, where the riches of sequence data will occupy many future generations of scientists, and in human genetics, where there are many families that, thanks to prenatal examinations, have healthy children. Perhaps, though, this usefulness will be limited in its validity by other reservations; a prenatal examination on the basis of

DNA is perhaps unacceptable to abortion opponents. The recording of DNA data could disconcert those who see a threat to privacy. It is encouraging that the National Health Institutes and the Department of Energy in the United States have made considerable financial means available to ensure that these issues are addressed and that not only specialists but also the public take part in the debates. Many scientific projects of large scope—the supercollider, the space station, "Star Wars" are regarded as unimportant at best and a catastrophe for humanity at worst. The various plans for the analysis of the human genome are intended in the end to improve the lives of many thousands of people. Is it naive to hope that we as a society will use the knowledge of our molecular genetic make-up in such a way that we feel our responsibilities and respect for others even more strongly?[2]

## WHY LIVE? WHY DIE?

Perhaps one day we will be able to live forever. Today, through technological progress and surgical intervention, we are able to gain control over our biology and tread a little closer to the oldest dream of humanity. If air, water, and land don't run out before these developments have reached full maturation, even newly built humans could continue to live—with or without us.

Perhaps we too find ourselves on the biological threshold as our ancestors once did, from proto-human to modern human. Such a metamorphosis of proto-humans occurred in history, and in a relatively recent instance, when the Neanderthals, who walked the earth for two hundred thousand years, breathed their last about thirty-three thousand years ago, possibly in a cave near Malaga, Spain. Since life in itself is increasingly interconnected overall, evolution will certainly not reverse its course. Proof of this lies with the numerous vestigial structures in the body that seem senseless today but are possessed by nearly all animals. They did not reverse or "evolve away" because they were already integrated into the structure of the body. For example, the absence of a skeleton in mollusks (snails, mussels, squid) prevents this group from ever attaining a dominant role on dry land. Insects and spiders (arthropods) have the same problem. Although they are, as far as sheer numbers go, the most successful organisms on the planet, they can never exceed a certain body size. Their thin exoskeleton would break apart if they exceeded a certain body mass.

Of course, insects, spiders, and mollusks are not necessarily at a disadvantage to vertebrates because of their different body structures. On the contrary, they are beautifully and inimitably adapted to their environments. The example of the skeleton is only meant to show how certain basic characteristics in animals limit their development in certain directions—in this case the development of larger species. Quite a boon for phobic types: a monster like the spider in the film *Arachnophobia* is impossible in real life because it would collapse and break apart (in addition, it would start to burn once it tried to move because it would produce too much heat energy).

There can be biological developments forward, but none backward. We humans cart around some useless body parts that used to have a function but have no hope of being "recalled," and serve only to remind us of our origins. Among such body parts are the appendix or wisdom teeth. We weren't even meant to walk entirely upright—our tendency to varicose veins and hernias is clear evidence of this. Even the pelvis should be structured differently: when we are born, our enlarged skull often no longer fits through the bony pelvic structure of the female human. As a result, a human newborn's skull has to grow rather quickly in the first weeks and months of life. "The human skeleton is far too long, the pelvic bones and the legs are too weak for the entire weight of the body—no manufacturer would dare bring a car with so many flaws onto the market," the scientist Hermann Römpp writes in his small book *The Future of the Earth and the Human*. All our bodily imperfections are relics from our four-legged past.

The organs mentioned (and a further eighty in the human body) slowly grow smaller or better adapted during the course of human development, although the appendix and the wisdom teeth will probably never disappear entirely. The reason for this is that they no longer represent any serious biological burden and do not put those who have them at any disadvantage to those who do not. Other physical traits, such as the structure of the pelvis, have adapted themselves in all their shortcomings as much as they possibly can to our current upright standing position. An alternate construction is as impossible as it is for insects and squid.

There are more than these evolutionary dead ends. Some organs provide evidence of greater adaptability and wider ranges of function. For example, the already existing hand and larynx muscles have improved to such a great extent that we can speak and grasp things with comfort, ease, and readiness.

It remains uncertain whether immortality is a possible development in human history. According to our current understanding of biology, it would hardly be imaginable that such a development could bring any adaptational advantage. On the contrary, as discussed and argued here, eternal life, for several biological and psychological reasons, would never work. It would hinder continuous regeneration of organisms and would nip any necessary adaptations to the environment in the bud. And it is precisely this regeneration and adaptation that is the central element of evolution—to accelerate it, nature "invented" sexuality, which ensures ever new combinations of genes, new combinations of qualities and traits, and thus the adaptation to ever new environmental conditions. If there were no such thing as death, evolution would come to a standstill. Immortality would mean the final death of the species.

Death is an inextricable component in the course of life. Given all human history, we still have difficulty facing this fact. Goethe considered it an absolute honor for people to exist between the poles of life and death. In his poem "Blessed Longing," he brings this notion to a point:

*And as long as you do not have it,*
*This Dying and Becoming,*
*You are but a gloomy guest*
*On the dark earth.*

Better words on the consequences of immortality have hardly been written—immortals would have to be, for all the reasons mentioned in this book, "gloomy guests" indeed. Brain transplants, which might represent a possibility of lengthening life, could result in severe physical and mental problems. Today, nerve cells from dead embryos are implanted in people with Parkinson's disease. Professor Hinderk Emrich, head of the division of Clinical Psychiatry at the medical school of Hanover, Germany, hits the nail on the head when he says, "It is not easy to cope with the feeling of having the cells of an unborn human in one's own head."

Another already mentioned result of immortality is the worsening problem of overpopulation. This was described by professor of comparative anatomy A. W. Nemilow at University of St. Petersburg. His remarks underline the biological importance of death.

> Death creates balance in nature. If there were no such thing as death, the earth would be swamped by the current of life, it would find its demise in its own floods, and every life would cease to be. Animals would multiply and cover the earth with a thick, teeming crust. There would be so many birds in flight that they would bump into one another and shield us from all sunlight. Our oceans would become a thick brew, teeming with wriggling fish and other sea creatures. Flies, butterflies, beetles, and other insects would cover the earth with such a thick layer that not even the highest mountain tops would be visible.
>
> If this does not occur, it is because nature provides us with the "benefactor" Death. Of the enormous number of creatures born, death sweeps away all that are superfluous and leaves only those alive that possess the potential for further development and perfection.[3]

This potential should never be obstructed for us.

Biology not only explains to us the meaning of death. There is also an insight handed down to us from ancient Greece that is as philosophical as it is practical. The priests of Eleusis were able to "look death in the face." Their insight was as valid then as it is now, "A life without fear of death—that is the life of the gods."

*NOTES*

*Chapter 1. Why Death Is Part of Life*

1. The first cells came into existence approximately three billion years ago, a fact supported by the discovery of single-cell fossils. In the time before that—at most a billions years—all that existed were proteins, the basic building blocks of life. If we consider the entire history of living cells on a time line scaled to one year, then humanlike creatures appeared on earth on December 31 at 3:15 P.M. Before then, in October of this "year," all the major groups of invertebrates existed. Around December 17, the dinosaurs went extinct.

2. H. Dekker, "Alexis Carrel and the Cultivation of the Tissue of Adult, Warm-Blooded Animals Outside the Organism," *Kosmos Handweiser für Naturfreunde* [The Naturalist's Handbook to the Cosmos] 10(1)(1913): 57.

3. Ibid., p. 61.

4. A. W. Nemilow, *Leben und Tod* [Life and Death] (Leipzig, 1927), p. 99.

5. The observation that cells know the age of the body they inhabit has been made not only on fibroblasts but also on lungs, skin, liver, artery walls, eye lenses, and T-cells (a specific cell of the immune system). In order to be certain, tissue from aborted fetuses up to cells of ninety year olds were used. It was confirmed that all kinds of cells—"even those separated from their original neighboring cells"—knew their age. Cells from young bodies completed up to sixty divisions, while tissue from older people divided more and more seldom, the older the person.

6. My calculation: a sugar cube weighs about 3 grams, which is $10^{22}$ sugar molecules. The oceans of the earth contain a total of $10^{21}$ quarts of water. When stirred, the sugar cube consists of ten molecules for every quart of sea water and thus two sugar molecules per cup.

From a lecture given in 1933 by the famous German chemist Otto Hahn, one of the discoverers of nuclear fission, we have another nice comparison that brings the world of neurotransmitters and particles to scale: "Imagine an ordinary light bulb. It has a vacuum. If we were to make even the tiniest hole in the bulb so that one million molecules of air per second were sucked into the vacuum, it would take more than 100 million years before the interior of the light bulb had the same air pressure (and the same number of air particles) as the rest of the earth outside."

*Chapter 2. No One Wants to Die*

1. R. Riedl, *Biologie der Erkenntnis. Die stammesgeschichtlichen Grundlagen der Vernunft* [Biology of Insight: The Species-Historical Foundation of Reason], 2d ed. (Berlin and Hamburg, 1980), p. 91 ff.

2. L. McLaughlin, "Two at Harvard Share Nobel in Medicine," *Boston Globe,* October 10, 1981.

3. K. Aram, *Magie und Mystik in Vergangheit und Gegenwart* [Magic and Mysticism in the Past and Present (1929)], from the most recent unabridged edition (Berlin, 1993), pp. 47–48.

4. Ibid., pp. 51–52.

5. Since 1990 I have come across only four more or less credibly confirmed cases of states of apparent death (*Scheintod*)—a truly small number, in spite of a comprehensive search. A contemporary overview of newspaper reports on this subject can be found in the July 1996 edition of the Scottish magazine *Fortean Times.* Further reports can be found in the book *Lebendig begraben* [Buried Alive] (Augsburg, 1995) by Berlin physicians Tankred Koch, with Otto Prokop.

6. From W. J. Schraml, *Einführung in die moderne Entwicklungspsychologie für Pädogogen und Sozialpädagogen* [Introduction to Modern Developmental Psychology for Pedagogues and Social Pedagogues] (Munich, 1972).

7. Ch. W. Hufeland, *Macrobiotik oder die Kunst sein Leben zu verlängern* [Macrobiotics; or, The Art of Prolonging Life] (Jena, 1800), pp. iii–v.

8. E. Haeckel, *Gott-Natur (Theophysis). Studien über monistische Religion* [God-Nature (Theophysis): Studies on Monistic Religion] (Leipzig, 1914), pp. 44–45; also in H. Schmidt, ed., *Gemeinverständliche Werke* [Generally Comprehensible Works of E. Haeckel] (Leipzig, 1924), 3:462–463.

9. Quoted by M. McCutcheon, *The Compass in Your Nose: And Other Astonishing Facts About Humans* (New York, 1989), p. 213 ff.

10. Several of these repair molecules were even named "molecule of the year" by *Science* in December 1994. "Every day," writes *Science,* "more than 10,000 DNA building blocks are lost by each cell of the body. Fortunately, DNA repair molecules are on the case. Like a perfectly practiced repair team, they look tirelessly for DNA errors, cut out defective parts and then fill the holes."

11. The "biblical" age as well as the calculated 350 years did not really happen in biblical times, as it is passed down to us in Scripture. As a result of contemporary theological belief, ages in the Hebrew Bible, which range from 70 to 969, are to be understood symbolically. According to Dr. Lamberty-Zielinski, of the Cologne diocese, they are intended to emphasize that "a long life and old age were gifts from God generally to those who lived according to His will (e.g., Moses, Noah, Job, among others)."

12. V. B. Dröscher, *Mich laust der Affe. "Fabelhafe" Redewendungen aus der Welt der Tiere* [Well, Blow Me Down! "Splendid" Expressions from the Animal Kingdom] (Düsseldorf, 1981), p. 67.

13. R. Prinzinger, *Das Geheimnis des Alterns. Die programmierte Lebenszeit bei Mensch, Tier und Pflanze* [The Secret of Aging: Programmed Lifespan in Humans, Animals, and Plants] (Frankfurt am Main, 1996), p. 446.

14. Quoted according to D. Summerlin, "No Smoking," *Nature* 371:113.

15. H. Franke, *Auf den Spuren der Langlebigkeit* [Seeking the Tracks of Longevity] (Stuttgart, 1985), pp. 83–84.

16. It must be added that a better education is usually made possible by better socioeconomic circumstances. At the same time, this means that children who enjoy a prolonged education generally receive better care and nourishment at home, which promotes health and also the attainable age. The connection could also be stated thus: socially well-positioned city children are healthier and, as a result, live longer. The extended school education is only a by-product of better living conditions and not the cause of a longer life.

17. Of course, milk bacteria are not unhealthy or without any effect. The single-celled organisms do not, however, extend life, as Elias Metschnikov and Loudon Douglas claimed. Experiments by Silvia Gonzalez from the Centro de Referencia para Lactoacillos in Chacabuco, Argentina have demonstrated that the milk bacteria *Lactobacillus casei* and *Lactobacillus acidophilus* can prevent an experimentally in-duced dysentery in mice. Both tested milk bacteria are sold, overpriced, in several kinds of yogurt. Mixtures of milk and commonplace intestinal bacteria, designed to influence the animals' digestive tract, are added to chicken feed on intensive live-stock farms. Kefir is still sold in every German supermarket.

18. An extensive overview of the scientific studies done by Pauling and his re-search institute can be found in a special library at Oregon State University. It can be accessed on the Internet at http://osu.orst.edu/dept/spc/

19. Only ovulation occurs periodically, approximately two weeks before men-struation. Menstruation itself, during which shreds of mucous membrane, tissue fluid, as well as blood and white blood cells are excreted from the body, can be slowed down or delayed by stress and many other influences, thus becoming irregu-lar. Before the mucous membrane is excreted, it breaks down in the uterus over the course of three to eight days. The speed of this process depends on numer-ous physical and external influences that are not subject to any predictable time frame.

20. H. Schlieper, *Das Raumjahr. Die Ordnung des lebendigen Stoffes* [The Space Year: The Order of Living Matter] (Jena, 1929).

21. Another example: A "23" also comes out when the date of WTC disaster is added: 11.9.2001 11 + 9 + 2 + 0 + 0 + 1 = 23. 23 is not only a central number in biorhythm but also in modern conspiration theory.

22. This compilation comes primarily from *Psychologie heute* [Psychology To-day], July 1992.

23. Long-term experiments of this kind are carried out today only by very few volunteers, usually under self-supervised conditions. The reason: during the return

to normal time, physical damage and disorientation after more than six months of being cut off, similar to an extreme form of jet lag.

### Chapter 3. The Immortality of the Individual: Possibilities and (for Today) Impossibilities

1. D. Linke, *Hirnverpflanzung. Die erste Unsterblichkeit auf Erden* [Brain Transplantation: First Immortality on Earth] (Reinbek, 1993).

2. J.-D. Bauby, *Schmetterling und Taucherglocke* [Butterfly and Diving Bell] (Vienna, 1997).

3. More on this at http://www.benecke.com/maggots.html

4. The theory of selfish genes can be found in R. Dawkins, *The Selfish Gene* (New York and Oxford: Oxford University Press, 1990).

5. In contrast to the embryonic traits adult humans carry, in the animal kingdom there are real cases of degeneration. For example, after one life as reproductive males, several centipedes are capable of returning to their final larva stage by sloughing their skins. These males are called "switch males" (*Schaltmännchen*) and can transform themselves with one more sloughing of the skin into sexually mature animals again. Since they grow additional legs with each sloughing, the switch males are longer than their normal fellow centipedes that cannot transform themselves. The zoologist K. W. Verhoeff reports that the life of these animals is lengthened by the sloughing process (known as "period morphosis") by two to six and a half years.

6. S. Blackmore, "Minds Possessed—Dark White: Aliens, Abductions, and the UFO Obsessions," *Nature* 372:290.

### Chapter 4. Humanity—Immortal?

1. James Lovelock, *The Ages of Gaia: A Biography of Our Living Earth* (New York: Norton, 1988).

2. It is a fact that the world population is growing. In the American World Population Profile from 1996, the expected average for the upcoming years is estimated as follows:

2000: 6.1 billion
2010: 6.9 billion
2020: 7.6 billion

The number of births per female will supposedly go down from 2.8 in 2000 to 2.3 by 2020.

Noteworthy in context is that the number of deaths per 1,000 is declining rapidly—fewer people are dying worldwide:

| Region | 1996 | | 2020 | |
| --- | --- | --- | --- | --- |
| | Inhabitants | Death rate | Inhabitants | Death rate |
| South Sahara and Africa | 594 | 103 | 1022 | 65 |
| Middle East and North Africa | 294 | 55 | 483 | 22 |
| China | 1230 | 30 | 1440 | 17 |
| The rest of Asia | 2070 | 55 | 2760 | 38 |
| Latin America and the Caribbean | 488 | 43 | 643 | 21 |
| Europe and the CIS | 799 | 43 | 833 | 25 |
| North America | 295 | 8 | 361 | 5 |

3. But things could turn out quite differently.

A fertilized egg cell divides very rapidly at first, but only a small number of cells form overall. Once a decent number of cell groups grow, development proceeds quite rapidly. A child develops from an embryo, an adolescent from a child, and a young adult from that child. Such growth can be observed. This corresponds to the "steep face" of very rapid growth and enormous weight gain. As soon as a person completes the "steep face" phase, things proceed more leisurely, until growth reaches a "plateau." Weight and cell numbers stay more or less constant. Like individuals, groups of organisms behave similarly, as in the intestinal bacteria so often used in investigations by biologists called *Escherichia coli*, or *E. coli* for short. When placed in a nutrient solution at 98.4° F (37° C), many bacteria develop in the first few hours. About every twenty minutes, each of the bacteria divides in the solution. We can then observe under a microscope every quarter-hour just how quickly the tiny bacteria continue to multiply. Suddenly we can count many more bacteria than before. The bacteria have conquered the initial growth phase and are now multiplying so rapidly that the growth curve rises sharply, like the "steep face." After a few more checks, calm returns, and the number of bacteria remains high but constant.

This kind of growth—slow climb, steep climb, plateau—appears in masses of bacteria as it does in individual humans. Isn't it possible that this principle also applies to the growth of all humanity? Perhaps we are now in the phase of rapid growth. If this were the case, the number of people on earth might level off at 8 or 9 billion.

4. In 1994, the Canadian biologist Douglas Morris tried to count all species living on earth and all those that were extinct. His result can be found in *Nature* 373:25: since 1600, 584 plant species and 485 animal species have gone extinct, while countless more are endangered. Increasing human expansion across the earth is the primary reason for the disappearance of other species. Regions inhabited by humans expand annually by about 1.7 percent; since 1700, jungles and forests have

shrunk to one-tenth of their original size. Within the last forty-five years, 17 percent of regions covered with plant life have been lost through overgrowth or clearance. About 9 million hectares of land are completely destroyed biologically, a further 30 million are on the brink.

5. V. Delage and M. Goldsmith, *Die Entwicklungstheorien* [Theories of Evolution] (Leipzig, 1913).

6. The species of Chinese Cochin China monkey was an immediate victim of defoliation caused by Agent Orange. In 1997, only about thirty of these delicate creatures still existed worldwide in zoological gardens and wildlife preserves, dependent on at least one meal a day from the leaves of trees. Back in 1968, a few of these primates were brought from Vietnam and placed in zoos, where, apparently, they did not thrive. If this species (*Pygathrix nemaeus nemaeus*) were to go extinct over the next few years, it would be the direct result of military action that took place nearly forty years ago.

7. For example, the colorful arctic tern was suddenly deprived of its most important food source, the herring. Starting in 1969, a half-million young arctic terns starved to death each year on the Lofoten Island Røst alone. The parent animals were able to get by on a lean diet of sprats and sand eels, but there was no hope for the birds' offspring. Only fourteen years later, when the herring population recovered, did arctic tern populations rebound. If a food source catastrophe were to strike again, arctic terns and other members of that bird family would have slim chances of survival. This is because the adult birds' alternate food source, the sand eel, is being fished increasingly.

The large-scale exploitation of the oceans first became possible because of several technical inventions in the fishing industry. Schools of herrings are located from afar with the aid of horizontal and vertical echo sounders. Deep dragging nets that, with the help of sonar, can be adjusted precisely to the depth at which the herring are swimming, capture the activity of the central school of fish.

The Russians learned another trick from dolphins. When a group of ocean mammals attacks a school of herring, they emit a loud hissing sound. With this sound, the dolphins imitate the language of the herrings. The hissing means nothing so much as "Everyone, school together!"

This noise is transmitted by Russian ships working in the canned fish industry. Before this technique was perfected, many herrings escaped. Now, when they hear the hissing, they group exactly where the fishermen want them: in the net.

But the worst method of industrial fishing takes the form of an invention from Iceland: the purse seine fishing net. With an enormous net that hangs like a curtain from buoys, an entire school is surrounded. Then the bottom end is tied together like a reversed sack. The number of fish caught by such a method is so tremendous that it can't be heaved on deck with one motion. Huge suction dredgers are pushed into the mass and pump the herrings through pipes into the storage room. From a school of millions, not a single herring can escape.

The Germans have also played their part in the high efficiency of fishing ships. The Integrated Fishing System promoted by the federal government starting in 1970 coupled modern methods of location with computer-aided analysis. Schools of fish appear graphically on the screen while a geographical course is calculated simultaneously. When, in 1978, the Integrated Fishing System was completed and implemented, the herring industry had already collapsed for the first time—the irony of fate.

If, in the next few years, industrial fishing continues on this course, we will witness an interesting experiment. It will be called, "Can the North Sea live without herrings?"

*Chapter 5. The Meaning of Life—"Biologically Speaking"*

1. Involved are, among others, the gene age-1, daf-2, clk-2, and unc-24. Cf. B. Lakowski and S. Hakimi, *The Road to Hell (or Heaven) Is Paved with Many Genes,* Worm Meeting abstract 341 (1997), University of Wisconsin, Madison.

2. J. D. Watson, M. Gilman, J. Witkowski, *Recombinant DNA,* 2d ed. (New York, 1992).

3. A. W. Nemilow, *Leben und Tod* [Life and Death] (Leipzig, 1927).

# SUGGESTED FURTHER READING

Alberts, Bruce et al. *Essential Cell Biology: An Introduction to the Molecular Biology of the Cell*. Garland, 1997.

Arkin, Robert. *Biology of Aging*. 2d ed. Sunderland, Mass.: Sinauer, 1998.

Benecke, Mark. "DNA Typing in Today's Forensic Medicine and Criminal Investigations: A Current Survey." *Naturwissenschaften* 84 (1997): 181–188.

Bishop, Jerry E. and Michael Waldholz. *Genome: The Story of the Most Astonishing Scientific Adventure of Our Time*. New York: Simon and Schuster, 1990.

Brown, Lester R. *State of the World, 1996: A Worldwatch Institute Report on Progress Toward a Sustainable Society*. New York: Norton, 1996.

Cairns-Smith, Alexander Graham. *Seven Clues to the Origin of Life: A Scientific Detective Story*. New York: Cambridge University Press, 1990.

Council of Environmental Quality/U.S. State Department, Gerald O. Barney, study director. *The Global 2000 Report to the President*. Washington: U.S. Printing Office, 1980.

Darwin, Charles Robert. *The Collected Papers of Charles Darwin*. Chicago: University of Chicago Press.

Dawkins, Richard. *River Out of Eden*. New York: Basic, 1995.

Dawkins, Richard. *The Blind Watchmaker: Why the Evidence of Evolution Reveals a Universe Without Design*. New York: Norton, 1996.

Dawkins, Richard. *The Selfish Gene*. Oxford: Oxford University Press, 1990.

de Kruif, Paul. *Microbe Hunters*. Fort Washington, Penn.: Harvest, 1996.

Dixon, Bernard. *Power Unseen: How Microbes Rule the World*. Oxford: Oxford University Press, 1995.

Eccles, Sir John C. and Daniel N. Robinson. *The Wonder of Being Human: Our Brain and Our Mind*. New York: Random House, 1986.

Eigen, Manfred and Ruthild Winkler-Oswatitsch. Trans. Paul Woolley. *Steps Towards Life: A Perspective on Evolution*. Oxford: Oxford University Press, 1992.

Eldredge, Niles. *Macroevolutionary Dynamics: Species, Niches, and Adaptive Peaks*. New York: McGraw-Hill, 1989.

Eldridge, Niles. *The Miner's Canary*. New York: Princeton University Press, 1994.

Gardner, John, ed. *Gilgamesh: Translated from the Sin-Leqi-Unninni Version*. Trans. John Maier. New York: Random House, 1985.

Gornitz, V., C. Rosenzweig, and D. Hilel. "Is Sea Level Rising or Falling?" *Nature* 371(1994): S. 481.

Heppener, Frank H. *Professor Farnsworth's Explanations in Biology*. New York: McGraw-Hill, 1990.

Huxley, J. *Evolution in Action*. London: Chatto and Windus, 1953.

Illich, Ivan and R. Mendelsohn. "Medical Ethics: A Call to Debunk Bioethics." In Ivan Illich, *In the Mirror of the Past: Lectures and Addresses, 1978–1990*. New York: Marion Boyars, 1992.

Leakey, Richard and Roger Lewin. *The Sixth Extinction: Patterns of Life and the Future of Humankind*. New York, Bantam, 1995.

Lovelock, James. *Gaia: A New Look at Life on Earth*. 3d ed. Oxford: Oxford University Press, 2000.

———— *The Ages of Gaia: A Biography of Our Living Earth*. New York: Norton, 1988.

Lovins, Amory B. *Soft Energy Paths: Towards a Durable Peace*. New York: Harper Collins.

McCutcheon, Marc. *The Compass in Your Nose and Other Astonishing Facts About Humans*. Los Angeles: Tarcher, 1989.

McKibben, Bill. *The End of Nature*. New York: Anchor, 1999.

Moody, Raymond A. and Kubler-Ross, Elisabeth. *Life After Life: The Investigation of a Phenomenon—Survival of Bodily Death*. 2d ed. San Francisco: Harper, 2001.

Nuland, Sherwin B. *How We Die: Reflections on Life's Final Chapter*. New York: Knopf, 1994.

Pearl, Raymond. *The Biology of Death*. Philadelphia: Lippincott, 1922.

Raup, David M. *Extinction: Bad Genes or Bad Luck?* New York: Norton, 1991.

Reiter, Russel J. and Jo Robinson. *Melatonin: Your Body's Natural Wonderdrug*. New York: Bantam, 1995.

Riedl, Rupert. *Biology of Knowledge: The Evolutionary Basis of Reason*. New York: Wiley.

Rose, Kenneth J. *The Body in Time*. New York: Wiley, 1988.

Rosenberg, Steven A. *The Transformed Cell Unlocking the Mysteries of Cancer*. New York: Putnam, 1992.

Rosenfield, Israel. *The Strange, Familiar, and Forgotten: An Anatomy of Consciousness*. New York: Vintage, 1993.

Shapiro, Robert. *The Human Blueprint : The Race to Unlock the Secrets of Our Genetic Script*. New York: St. Martin's, 1991.

Singer, Peter. *Practical Ethics*. 2d ed. New York: Cambridge University Press, 1993.

Thompson, Larry. *Correcting the Code: Inventing the Genetic Cure for the Human Body*. New York: Simon and Schuster, 1994.

Thompson, Richard F. *The Brain: A Neuroscience Primer*. 2d ed. New York: Free-man, 1993.

United States Congress Joint Economic Committee, Subcommittee on International Economics. The Global 2000 Report: Hearing Before the Subcommittee on International Economics of the Joint Economic Committee, Congress of the United States, Ninety-sixth Congress, second session, September 4, 1980. Washington, D.C.: U.S. Government Print Office.

USA Agency for International Development. *World Population Profile*. Washington D.C., 1996.

Vester, Frederic. *Unsere Welt, ein vernetztes System: e. internationale Wanderausstel-lung*. Stuttgart: Klett-Cotta, 1983.

Weiner, Jonathan. *The Next One Hundred Years: Shaping the Fate of Our Living Earth*. New York: Bantam, 1990.

Wilson, Edward O. and F. M. Peter. *Biodiversity*. Washington, D.C.: National Academy Press, 1988.

# INDEX

Page numbers for illustrations are in *italic*

Morris, Edwin, 129
Moynter, Jane, *161*
MRI (magnetic resonance imaging), 102
Mullis, Kary, 118–19, 122
Mummification: Egyptian, 39–40; Egyptian mummy can, *40*; retention of DNA, 40, 121
Mummy powder, 40, *40*
Munggona, New Guinea, 41
Murray, Joseph, 96
Museum of Hygiene (East Berlin), 55–56
Mustard weed (*Arabidopsis*), 165
Mutations, 9, 31

Nachtmin, 39
Neanderthals, 169
Near-death experiences, 33–34, 36
Nelson, Mark, *161, 162*
Nemilow, A. W., 171
Nemilow, Antoni, 12–13
Neocortex, 101, 103
Neoteny, 104–7
Neo-vitalists, 54
Neuroprosthesis, 140
New Guinea tree burials, *41,* 41–42
Niehans, Paul, 132
Ninevah, 37
Nucleotides, 2
Nutrition: caloric restriction, 61, 82; decreasing needs in aging, 59; life-lengthening, 74–77; macrobiotic diet, 53–54; Pauling recommendations, 79–80; prison food, 81; requirements, 47–48

Ocular dominance columns, 35
*On the Origin of Species* (Darwin), 54
Organic farming, 154
Organs, artificial, 138–39
Organ transplantation: donor rights, 134–35, 137–38; ethical issues, 134–36; harvest of organs, 101–2;

history of, 12; issues in, 133; legislation, 134; nonbiological problems of, 136–38; ownership of organs, 103, 133–35
Orians, Gordon, 150–51
Osborne, Thomas, 75
Osiris, 39
Osteoporosis, 49
Out-of-body experiences, 33–34
Ovaries, 51
Overfertilization, 154
Overpopulation, 145–46, 171
Oxygen, normal atmosphere, 142
Oxygen consumption, 61, 63–64
Oxygen supply, brain death and, 102

Pääbo, Svante, 120–22
Pair bonding, 90
*Pandanus adinobotrys,* 42
Paracelsus, Philippus Theophrastus, 10
Parkinson's disease, 125, 131, 171
*Paruoctonum mesaenis* (desert scorpion), 52
Pasteur, Louis, 87
Pauling, Linus, 77–80, *78,* 82, 175*n*18
PCR (polymerase chain reaction), 118–19
Pearl, Raymond, 12
People, vs. humans, 137
Perception, 128–29, 130–31
Period metamorphosis, 176*n*5
Perl, Thomas, 19
Personality, 116
Pescod-Taylor, David, 44
Pesticides, 154
Pets, positive effects of, 59
Pfeiffer, E. W., 150–51
Pfister, Christian, 147
Pfurtscheller, Gerd, 100
P53 gene, 18
Physical fitness, nutrition and, 74
Pierpaoli, Walter, 82
Pinworm, *see Caenorhabditis elegans*
Pius XII, Pope, 132